Solution Synthesis of Inorganic Films and Nanostructured Materials

MATERIALS RESEARCH SOCIETY
SYMPOSIUM PROCEEDINGS VOLUME 1449

Solution Synthesis of Inorganic Films and Nanostructured Materials

Symposium held April 9–13, 2012, San Francisco, California, U.S.A.

EDITORS

Menka Jain

University of Connecticut
Storrs, Connecticut, U.S.A.

Xavier Obradors

Institut de Ciència de Materials de Barcelona, CSIC
Catalunya, Spain

Quanxi Jia

Los Alamos National Laboratory
Los Alamos, New Mexico, U.S.A.

Robert W. Schwartz

University of Missouri System
Columbia, Missouri, U.S.A.

Materials Research Society
Warrendale, Pennsylvania

CAMBRIDGE
UNIVERSITY PRESS

Shaftesbury Road, Cambridge CB2 8EA, United Kingdom

One Liberty Plaza, 20th Floor, New York, NY 10006, USA

477 Williamstown Road, Port Melbourne, VIC 3207, Australia

314–321, 3rd Floor, Plot 3, Splendor Forum, Jasola District Centre, New Delhi – 110025, India

103 Penang Road, #05–06/07, Visioncrest Commercial, Singapore 238467

Cambridge University Press is part of Cambridge University Press & Assessment, a department of the University of Cambridge.

We share the University's mission to contribute to society through the pursuit of education, learning and research at the highest international levels of excellence.

www.cambridge.org
Information on this title: www.cambridge.org/9781605114262

Materials Research Society
506 Keystone Drive, Warrendale, PA 15086, USA
http://www.mrs.org

First published 2012

CODEN: MRSPDH

A catalogue record for this publication is available from the British Library

ISBN   978-1-605-11426-2   Hardback

CONTENTS

SOLUTION SYNTHESIS OF METAL-OXIDE FILMS

*Invited Paper

v

NANOSTRUCTURES, NANORODS, AND SOLAR OR GAS SENSING APPLICATIONS

NANOSTRUCTURES AND NANOCOMPOSITE FILMS

THIN FILMS, CERAMICS, NANOPARTICLES, AND APPLICATIONS

PREFACE

Symposium BB, "Solution Synthesis of Inorganic Films and Nanostructured Materials" was held during the 2012 MRS Spring Meeting in San Francisco, California, on April 9–13, 2012.

In recent years significant progress has been made in synthesis of advanced functional materials using chemical solution routes. This symposium was focused on solution synthesis approaches for the growth of a wide-range of advanced functional inorganic thin film and nanostructured materials. During this symposium, developments in synthetic approaches of inorganic functional materials to achieve enhanced and/or novel functionalities for a variety of applications were highlighted.

Recent results were presented on the growth of: (i) highly crystalline, nano-patterned and composite functional oxide films, (ii) nanoparticles and nanocrystals, and (iii) self-assembled nanostructures by various chemical solution methods. A strong increased interest in low-cost and high throughput synthesis of functional and multifunctional inorganic materials indicates the worldwide importance of such synthetic methods. The symposium promoted information exchange between worldwide researchers from universities and national labs and engineers from industry. Various applications of solution grown inorganic materials were discussed that include gas sensing, photovoltaic, optical, plasmonics, memory devices, spintronics, bio-medical, superconducting, and magnetic-field sensing.

At this symposium, 191 papers were presented and more than 100 attendees were present at many of the sessions. Oral presentations covered four days and poster sessions were held on three evenings. The papers in this proceedings volume provide a glimpse of the recent developments in the chemical solution growth of nanoparticles, nanocrystals, films, and nanostructured materials for various applications.

Menka Jain
Xavier Obradors
Quanxi Jia
Robert W. Schwartz

July 2012

ix

ACKNOWLEDGMENTS

The papers published in this volume result from the MRS Spring 2012 Symposium BB. We sincerely thank all of the oral and poster presenters of the symposium who contributed to this proceedings volume. We also thank the reviewers of these manuscripts, who provided valuable feedback to the editors and authors. We greatly appreciate MRS publication staff for their constant help and for guiding us smoothly through the submission/review/decision process.

MATERIALS RESEARCH SOCIETY SYMPOSIUM PROCEEDINGS

MATERIALS RESEARCH SOCIETY SYMPOSIUM PROCEEDINGS

Prior Materials Research Society Symposium Proceedings available by contacting Materials Research Society

Solution Synthesis of Metal-Oxide Films

Mater. Res. Soc. Symp. Proc. Vol. 1449 © 2012 Materials Research Society
DOI: 10.1557/opl.2012.918

Pulsed Laser Assisted Polycrystalline Growth of Oxide Thin Films for Efficient Processing

Tomohiko Nakajima, Kentaro Shinoda and Tetsuo Tsuchiya
Flexible Chemical Coating Group, Advanced Manufacturing Research Institute, National Institute of Advanced Industrial Science and Technology, Tsukuba Central 5, 1-1-1 Higashi, Tsukuba, Ibaraki, 305-8565, Japan

ABSTRACT

We have investigated the polycrystalline growth by means of an excimer laser assisted metal organic deposition process and the strategy for the efficient growth. It was revealed that the pulsed photo thermal heating properties must be controlled by changing the laser fluence according to the substrate properties, such as thermal diffusivity. The threshold of the t_{eff} value for initial crystal nucleation is approximately 70 ns for oxide thin films. For the fabrication of good quality films with high crystallinity and without a laser ablation of the film surface, it is necessary that the irradiated laser fluence is adjusted to the conditions of t_{eff} (efficient annealing time) > 70 ns and T_{max} (maximum temperature) < T_{m} (melting point). Obtained oxide films by using the pulsed UV laser has large crystallite size, and it well functioned to enhance physical properties of films. For further efficient growth for polycrystalline growth of the oxide films, the starting solution containing nanoparticles is very useful: it is named as photo-reaction of nanoparticles process.

INTRODUCTION

Numerous inorganic materials which have been investigated for future electronics devices are usually incorporated in the devices as thin films. There are two main requirements for film growth methods for future applications: (1) Efficient fabrication for lower cost, and (2) low temperature operation for substrates which cannot withstand high temperatures (T > 500 °C). It is now widely accepted that chemical solution deposition (CSD) methods, such as a metal organic deposition (MOD) [1,2] and sol-gel [3,4] processes are better than physical and chemical vapor deposition processes in terms of process simplicity and low cost. However, the application of CSD as a conventional heating process is also an issue for low temperature fabrication, and is generally limited by the heat resistance properties of the substrate. To resolve this problem, we have developed processes by means of a combination of CSD and ultraviolet (UV) irradiation [5–8]. We have investigated a new method to fabricate oxide thin films at low temperatures using an advanced conventional MOD process, wherein oxide films are crystallized by means of excimer laser irradiation (λ = 193, 248, and 308 nm) instead of high temperature furnace heating in the MOD process [9]. We have named this process as an excimer laser-assisted metal organic deposition (ELAMOD).

Thus far, we have demonstrated that not only polycrystalline films but also epitaxial films of oxide materials can be grown by ELAMOD, when a single crystal substrate has a small lattice mismatch with the film [9,10]. In these studies, we noticed that crystal growth of oxide films under UV laser irradiation in the CSD process cannot be interpreted by only a photo-thermal heating effect [7,10]: some photo-chemical reaction at growth interface should be taken into account to the crystal growth mechanism in this process. Concerning the polycrystalline growth of oxide thin films, we mainly focused on the controlling parameters which affect an initial nucleation on amorphous and/or polycrystalline substrates [11]. In this paper, we investigate that key techniques for efficient polycrystalline growth under pulsed UV laser irradiation in the ELAMOD and recent developed photo-reaction of nanoparticles (PRNP) processes [12,13].

EXPERIMENT

For preparations of polycrystalline oxide thin films by means of the ELAMOD in this study, the starting solution was firstly prepared by mixing solutions of fatty acid salts dissolved in organic solvents including the constituent metals to obtain the required concentration and viscosity for spin coating. The prepared solution was spin-coated onto amorphous glass substrates at 4000 rpm for 10 s. The coated films were dried at 100 °C in air to remove the solvent, and then heated to 400 °C in air for 10 min to decompose the organic components of the films. The preheated films were irradiated with a KrF laser using a Compex110 Lambda Physik at pulse duration of 25 ns in air. The laser energy was homogenized by a beam homogenizer in 7 x 7 mm^2 area. The structural properties of the films were studied by a SmartLab RIGAKU X-ray diffractometer with Cu K_α radiation, V = 40 kV, and I = 30 mA. The cross-sectional transmittance electron microscopy (XTEM) was carried out using a JEM-2010 JEOL instrument operating at 200 kV. Temperature variations during the laser irradiation process can be described by the heat diffusion equation simplified into a one-dimensional heat flow as follows [10]:

$$\rho C \frac{\partial T}{\partial t} = \kappa \frac{\partial^2 T}{\partial z^2} + \alpha I(z,t)$$

where T is the temperature function of time t and depth z, ρ is the mass density, C is the specific heat capacity, α is the optical absorption coefficient, κ is the thermal conductivity, and the $I(z,t)$ is the laser power density. The laser power $I(z,t)$ is given by:

$$I(z,t) = I_0(t) \cdot (1 - R) \cdot \exp(-\alpha z)$$

where R is the reflectivity, and $I_0(t)$ was described as a smooth pulse approximated by:

$$I_0(t) = I_0 \cdot \left(\frac{t}{\tau}\right)^\beta \cdot \exp\left(\beta\left(1 - \frac{t}{\tau}\right)\right)$$

where I_0 is the incident pulse power density, τ_0 is the pulse duration, and β determines the temporal pulse shape. We carried out numerical simulations for the temperature variations for the excimer laser annealing process by means of a difference approximation based on the above mentioned equations.

DISCUSSION

For searching good conditions to make polycrystalline thin films in the ELAMOD process, we have to take into account photo-thermal heating profiles for first nucleation in the precursor matrix. Firstly, we can evaluate photo-thermal heating under a UV laser pulse by using the one dimensional heat flow model described above. Figure 1 shows the representative pulsed temperature rise and decay curve under the UV laser irradiation.

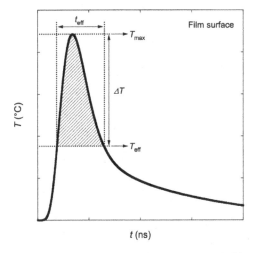

Figure 1. Schematic simulated temperature variation curve at precursor film surface under a UV laser irradiation by using a one dimensional heat flow. T_{max}, T_{eff}, and t_{eff} represent maximum temperature, effective temperature for crystallization and effective time for crystallization, respectively. The T_{eff} is a temperature in which the x-ray diffraction intensity can be detectable after furnace heating for 1h, this means nucleated nanodomains successfully emerged in the matrix. The t_{eff} is a time over the T_{eff}. The ΔT equals to $T_{max}-T_{eff}$.

Optimum conditions of laser irradiation are firstly chosen by keeping t_{eff} above 50–70 ns. This value has been obtained by precise control of t_{eff} and ΔT in the preparation of perovskite manganite thin films on glass substrates. The contour map for x-ray diffraction intensity of LaMnO$_3$ thin films after the KrF laser irradiation to the precursor with 1000 pulses at room temperature is shown in Fig. 2. The t_{eff} and ΔT were controlled by changing thermal diffusion property. The detailed conditions are shown in the previous literature [11].

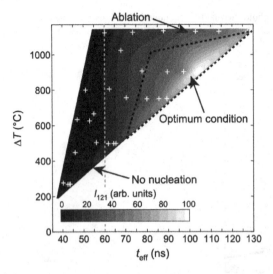

Figure 2. Contour map for the t_{eff} and ΔT dependence of the 121 peak intensities of LaMnO₃ thin films. White crosses indicate the observed points. T_{max} and T_{eff} in this system were 1920 °C and 790 °C, respectively [11].

As seen in this figure, there is clear threshold at around t_{eff} = 60 ns for the nucleation of LaMnO₃ nanodomains in the precursor matrix. Best condition for the polycrystalline growth was located above t_{eff} = 110 ns and at around ΔT = 850 °C. Above ΔT = 1000 °C, the crystallinity was reduced because the laser ablation would occur near melting point (T_m) of LaMnO₃ at 1900 °C. Thus, this result leads to a following conclusion: for an efficient polycrystalline growth from the amorphous LaMnO₃ matrix, it was revealed that the pulsed photothermal heating properties must be controlled by changing the laser fluence according to the substrate properties, such as thermal diffusivity. The threshold of the t_{eff} value for initial crystal nucleation is approximately 60 ns for this material. For the fabrication of good quality films with high crystallinity and without a laser ablation of the film surface, it is necessary that the laser fluence is adjusted to the conditions of t_{eff} > 90 ns and 500 °C < ΔT < 1000 °C. About the other materials, optimum t_{eff} values for the RbLaNb₂O₇ and CaTiO₃ were determined as *c.a.* 160 ns and 200 ns, respectively [14,15]. The thresholds of t_{eff} for nucleation of these materials were *c.a.* 70 ns. Therefore, we thought that these key parameters have universality to some extent in many oxide materials.

The notable feature of the polycrystalline growth under the laser irradiation is that crystallite size significantly increases if the laser irradiation conditions are optimized. This characteristic is linked to the epitaxial growth picture in the ELAMOD process. Under UV laser irradiation, the materials can be epitaxially grown on the crystal seed if it has similar crystal lattices with objective materials: it is strongly enhanced by photo-chemical effect at the reaction interface. In polycrystalline growth under the pulsed UV laser, first nucleation sites emerged by photo-thermal heating act as a seed, followed by a rapid succeeding crystal growth homoepitaxially from them enhanced by the abovementioned photo-chemical effect. Finally, the laser irradiation process can prepare highly crystalline oxide thin films with large grain size.

Figure 3(a) shows a XTEM image for a CaTiO$_3$ polycrystalline thin film on a glass substrate prepared by the ELAMOD process using a KrF irradiation at a fluence of 100 mJ/cm^2 for 5000 pulses and at 100 °C in air. The film thickness was evaluated to be c.a. 100 nm. Noteworthy is that it had very large grain size compared to the films which are usually obtained by conventional heating process: it was a single crystal in out-of-plane direction and has around 700 nm grains along the in-plane direction. Figures 3(b) and 3(c) show high-resolution XTEM images for each crystal grain. Both images clearly exhibit the grains are single crystal along the depth direction. The inset of Fig. 3(b) is the selected area electron diffraction (SAED) at the whole region of one grain shown in the Fig. 3(b). The SAED pattern was derived only from [111]-plane, indicating that it was evidently single crystal.

(a)

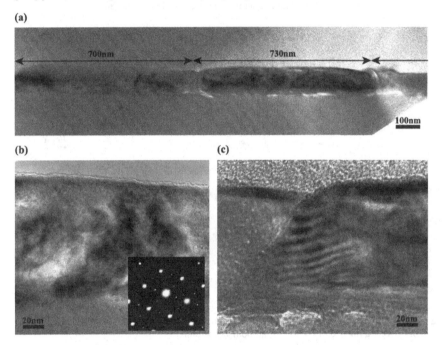

Figure 3. (a) Bright field image for the CaTiO$_3$ thin film. Arrows indicate each grain. (b,c) High resolution XTEM images. The inset of (b) shows the SAED pattern collected from the whole region of the grain.

Based on the polycrystalline growth picture and the characteristics in the ELAMOD process, we focus on a method for an efficient polycrystalline growth by using the pulsed UV laser. It is a use of nanoparticles as a seed into a starting solution. Actually, the first nucleation from an amorphous precursor matrix under the pulsed UV laser usually needs more than 1000 pulses. If a starting solution contains objective nanoparticles, the continuing crystal growth would rapidly occur. By means of this method called as PRNP, we have already efficiently prepared oxide thin films such as ITO and WO_3 [12,13]. For the preparation of WO_3 thin films, we used an alcohol solution contains WO_3 nanoparticles with the particle size at around 5–30 nm as shown in Fig. 4(a). The starting solution was deposited onto glass substrates by spin coating, then the particles coated substrate was irradiated by KrF laser at a fluence of 50 mJ/cm^2 for 7500 pulses at room temperature. For a comparison, precursor film was also treated by furnace heating at 500 °C for 60 min in air. Figure 4(b) shows the XTEM for the heat treatment sample. The crystallite size was similar with original particles. Conversely, it was much increased especially at near film surface prepared by the laser irradiation as shown in Fig. 4(c). The crystallites near the film surface were significantly larger, up to 70 nm, and their size gradually decreased as the depth increased. This is due to photo-thermal temperature decay from the surface.

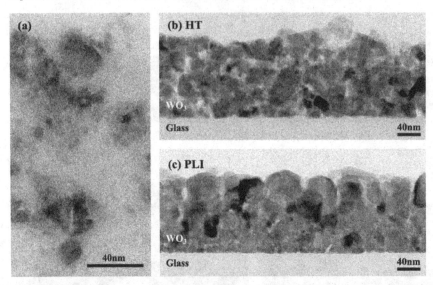

Figure 4. (a) Original WO_3 nanoparticles in the starting solution for the PRNP process. The prepared WO_3 thin films by means of (b) furnace heating (HT) at 500 °C and (c) pulsed laser irradiation (PLI) at a fluence of 50 mJ/cm^2 at room temperature.

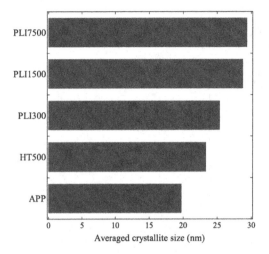

Figure 5. Calculated averaged crystallite size for the WO_3 as-prepared particles (APP) and thin films by heat treatment at 500 °C (HT500) and PRNP (PLI300, PLI1500 and PLI7500).

The averaged crystallite size of the samples prepared by heat treatment at 500 °C (HT500) and PRNP for 300 pulses (PLI300), 1500 pulses (PLI1500) and 7500 pulses (PLI7500) were evaluated by X-ray diffraction using Scherrer's equation (Fig. 5) [16]. By only 300 pulses irradiation, the averaged crystallite size exceeded that of heat treatment one. Then, it was saturated at around 1500 pulses. This indicates that the necessary pulse number for sufficient polycrystalline growth can be much reduced by introducing nanoparticles into a starting solution, leading a further cost-friendly process.

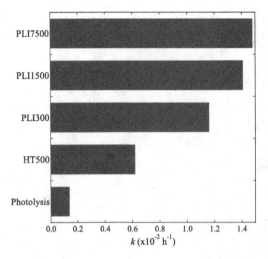

Figure 6. The rate constant for the photocatalytic degradation of methylene blue for the WO₃ thin films by heat treatment at 500 °C (HT500) and PRNP (PLI300, PLI1500 and PLI7500). Photolysis means a blank measurement.

The large crystallite growth under the pulsed UV laser irradiation as shown in Figs. 4 and 5 would be good for the physical properties of the films, which is especially related to electron transfer. Figure 6 shows the photocatalytic degradation rate constant (k) of methylene blue in aqueous solution for the prepared WO₃ thin films in which the surface was grafted by Cu(II) for enhancement of photocatalytic activity under visible light irradiation using Xe lamp at 100 kLux through a UV cut-off filter ($\lambda > 400$ nm). The k value was clearly increased by laser irradiation compared to the sample prepared by heat treatment. The surface crystallinity which depends on sufficient crystallite growth in WO₃ thin films was greatly improved by pulsed laser irradiation in the PRNP process. The enhanced crystallinity would stimulate the photoexcited electron/hole transfer at the film surface, resulting in the improvement of the photocatalytic activity. This kind of origin was also observed a transparent conducting material ITO prepared by the PRNP process, suggesting that the enhanced crystallite growth by means of the pulsed laser irradiation is efficient for enhanced physical properties of polycrystalline oxide thin films.

CONCLUSIONS

We have investigated the polycrystalline growth by means of the ELAMOD process and the strategy for the efficient growth. It was revealed that the pulsed photo thermal heating properties must be controlled by changing the laser fluence according to the substrate properties, such as thermal diffusivity. The threshold of the t_{eff} value for initial crystal nucleation is approximately 70 ns for oxide thin films. For the fabrication of good quality films with high crystallinity and without a laser ablation of the film surface, it is necessary that the irradiated laser fluence is adjusted to the conditions of $t_{eff} > 70$ ns and $T_{max} < T_m$. Obtained oxide films by using the pulsed UV laser has large crystallite size, and it well functioned to enhance physical properties of films. For further efficient growth for polycrystalline growth of the oxide films, the starting solution containing nanoparticles enables to stimulate the growth rate: it is named as PRNP process.

REFERENCES

1. A. C. Westerheim, P. C. Mcintyre, S. N. Basu, D. Bhatt, L. S. Yujahnes, A. C. Anderson and M. J. Cima, J. Electron. Mater. **22**, 1113 (1993).
2. T. Mihara, H. Yoshimori, H. Watanabe and C. A. P. Araujo, Jpn. J. Appl. Phys. **34**, 5233 (1995).
3. D. Avnir, V. R. Kaufman and R. Reisfeld, J. Non-cryst. Solids **74**, 395 (1985).
4. C. J. Brinker, A. J. Hurd and G. C. Frye, J. Non-cryst. Solids **121**, 294 (1990).
5. T. Nagase, T. Ooie and J. Sakakibara, Thin Solid Films **357**, 151 (1999).
6. T. Tsuchiya, A. Watanabe, Y. Imai, H. Niino, I. Yamaguchi, T. Manabe, T. Kumagai and S. Mizuta, Jpn. J. Appl. Phys. **38**, L823 (1999).
7. S. C. Lai, H-T. Lue, K. Y. Hsieh, S. L. Lung, R. Liu, T. B. Wu, P. P. Donohue and P. Rumsby, J. Appl. Phys. **96**, 2779 (2004).
8. C. S. Sandu, V. S. Teodorescu, C. Ghica, B. Canut, M. G. Blanchin, J. A. Roger, A. Brioude, T. Bret, P. Hoffman and C. Garapon, Appl. Surf. Sci. **208–209**, 382 (2003).
9. T. Tsuchiya, T. Yoshitake, Y. Shimakawa, I. Yamaguchi, T. Manabe, T. Kumagai, Y. Kubo and S. Mizuta, Jpn. J. Appl. Phys. **42**, L956 (2003).
10. T. Nakajima, T. Tsuchiya, M. Ichihara, H. Nagai and T. Kumagai, Chem. Mater. **20**, 7344 (2008).
11. T. Nakajima, T. Tsuchiya, M. Ichihara, H. Nagai and T. Kumagai, Appl. Phys. Express **2**, 023001 (2009).
12. T. Tsuchiya, F. Yamaguchi, I. Morimoto, T. Nakajima and T. Kumagai, Appl. Phys. A **99**, 745 (2010).
13. T. Nakajima, T. Kitamura and T. Tsuchiya, Appl. Catal. B **108–109**, 47 (2011).
14. T. Nakajima, T. Tsuchiya and T. Kumagai, Appl. Phys. A **93**, 51 (2008).
15. T. Nakajima, T. Tsuchiya and T. Kumagai, Cryst. Growth Des. **10**, 4861 (2010).
16. L. V. Azaroff, Elements of X-ray Crystallography, McGrawHill, New York (1968).

Mater. Res. Soc. Symp. Proc. Vol. 1449 © 2012 Materials Research Society
DOI: 10.1557/opl.2012.919

Can We Trust on the Thermal Analysis of Metal Organic Powders for thin film preparation?

Jordi Farjas[1], Daniel Sanchez-Rodriguez[1], Hichem Eloussifi[1,3], Raul Cruz Hidalgo[1,4], Pere Roura[1], Susagna Ricart[2], Teresa Puig[2], Xavier Obradors[2]

[1]Department of Physics, University of Girona, Girona, Catalonia, Spain.

[2]Institut de Ciència de Materials de Barcelona, CSIC, Bellaterra, Catalonia, Spain.

[3]Laboratoire de Chimie Inorganique, Faculté des Sciences de Sfax, Route de Soukra Km 3.5 BP 1171, 3000 Sfax, Tunisia

[4]Departamento de Física y Matemática Aplicada, Universidad de Navarra, 31080 Pamplona, Spain.

ABSTRACT

Thermal analysis techniques are routinely applied to characterize the thermal behavior of metal organic precursors used for oxide film preparation. Since the mass of films is very low, researchers do their thermal analyses on powders and consider that the results are representative of films. We will show here that, in general, this assumption is not true. Several examples involving precursors of $YBa_2Cu_3O_{7-x}$ (Ba and Y trifluoroacetates and Ba propionate) will serve to appreciate that films can behave very differently than powders due to their enhanced heat and mass transport paths. Ultimately, we will demonstrate that, in some cases, relying on powders thermal analysis may lead to erroneous conclusions.

INTRODUCTION

Metal organic precursors are widely used for the production of functional oxide thin films. After spreading and evaporating a solution containing the precursor salt, the film is pyrolyzed to remove the organic ligands, leaving the desired oxide film on the substrate [1]. The process parameters (heating rate, temperature, atmosphere) rely on the information delivered by thermal analysis experiments (notably, thermogravimetry and mass spectrometry) done on precursor powders. In fact, a survey of the literature devoted to chemical solution deposition (CSD) of thin films reveals that most authors perform thermal analysis on powders and that direct analysis on films is very scarce.

In this paper, we will give a number of examples showing that the thermal behaviour of powders may differ greatly from that of films. This is so because thermal decomposition of oxide precursors is, in fact, a solid-gas reaction and, consequently, the transport of gaseous reactants or volatile by-products plays a crucial role on the reaction kinetics [2]. Since gas transport is easier in films, their decomposition is faster than in powders. This means that the decomposition temperature will be lower in films. In case where decomposition is highly exothermic, heat dissipation can be of great concern. Whereas thermalization of thin films with the substrate is almost ensured, powders can experience large overheating that, in some circumstances, will lead to self-combustion. In the following, it will be shown that self-combustion is very usual in powders whereas it is very difficult (or impossible) for films thus contradicting the recent claim that oxide thin films can be obtained via self-combustion [3]. Finally, it must be emphasized that the reduced gas exchange in precursor powders may hidden the actual reaction mechanism operating in films and may lead to the synthesis of unwanted intermediate products (such as fluorides or carbonates).

EXPERIMENT

Trifluoroacetates (TFA) have been used in the form of commercially available powders whereas Ba propionate has been obtained from Ba acetate. Films of several hundreds of nm where analyzed.

Thermal analysis experiments were carried out with a Mettler Toledo TGA851^{e}LF and Setaram Setsys thermobalances that deliver simultaneously the differential thermal analysis (DTA) signal corresponding to the heat exchanged. Since, at the typical decomposition temperatures (200-400°C), the time response of the DTA signal is very long for the Mettler thermobalance (tens of s [4]), the DTA peaks have been deconvoluted by the apparatus time constant to obtain a DTA signal dependence on time (or temperature) closer to reality. A typical experiment consisted of heating the sample at constant rate (20 K/min) up to the maximum temperature. The TGA/DTA curves thus measured were then corrected by the apparatus baseline that was usually measured with a second heating ramp. Although this correction is essential for films, it can be sometimes skipped for powders. High purity gases at a flow rate around 50 mL/min were used to control the furnace atmosphere. The mass spectroscopy (MS) measurements were taken with an MKS Spectra Quadrupole (Micro Vision Plus), which detects molecular fragments with m/z < 300 amu.

XRD experiments were done in a D8 ADVANCE diffractometer from Bruker AXS.

RESULTS AND DISCUSSION

We will present several particular examples that illustrate the different thermal behavior of powders and films.

Transport of reactive gas

When one analyzes by TG the decomposition of Y(TFA)$_3$ powders, one realizes that the process to the final product (Y$_2$O$_3$) proceeds through several intermediate products (YF$_3$, Y$_6$O$_5$F$_8$, YOF) [5] and that, apart from minor differences in the mass-loss steps, no significant dependence on the furnace atmosphere (from Ar to wet O$_2$) is observed. The TG curves of powders decomposition in dry and wet O$_2$ are shown in Fig. 1. Until the discontinuity of the mass-loss derivative at 300°C in dry O$_2$ (whose origin will be explained in next section), both curves coincide. In addition we have verified that the main decomposition step (Y(TFA)$_3$ ---> YF$_3$) is also independent of the powder mass within the range analyzed (40 mg> m > 2 mg) [5].

From these results, one would expect the same behavior for Y(TFA)$_3$ films. However, this is not the case. Fig. 1 clearly shows that films decompose at a temperature lower than powders and that this shift is higher in wet air. So, in contrast with our conclusion drawn from powders, the experiments on films indicate that H$_2$O molecules play an important role in triggering the decomposition of Y(TFA)$_3$. It does not only participate in secondary reactions with volatile byproducts as pointed out in our precious study on powders [5].

14

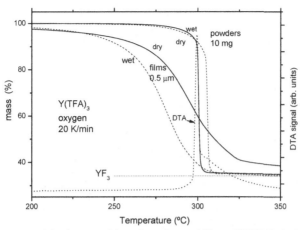

Figure 1.- TG curves of the decomposition of powders and films of Y(TFA)$_3$ in dry and wet O$_2$. The DTA curve in wet air for powders is also shown.

Although the detailed mechanism explaining the dependence on the H$_2$O partial pressure, P$_{H2O}$, in films is still an open question, we can understand very easily why this dependence is absent in powders. The lower surface to volume ratio of powders implies that H$_2$O transport to the sample from the furnace atmosphere is more difficult. Consequently, the H$_2$O consumption by the decomposition reaction produces a local depletion of H$_2$O (diminution of P$_{H2O}$) around the sample. In fact, in view of the curves of powders in Fig. 1 one must conclude that the local atmosphere near the sample is dry, even if wet O$_2$ is flowing through the furnace.

A similar explanation applies to the TG decomposition curves of Ba(TFA)$_2$ (Fig. 2). In this case, the decomposition temperature of 10 mg of powders is lower in air than in Ar, indicating that oxygen reacts with the precursor molecules. Since transport of O$_2$ molecules from the surface atmosphere to the sample surface is not instantaneous, this reaction necessarily diminishes the oxygen partial pressure, P$_{O2}$, near the sample and, owing to the higher surface to volume ratio of films, this effect is less pronounced in films. Therefore films will decompose quicker (at a lower temperature) because the local P$_{O2}$ will be higher than in powders.

Let us emphasize that, in these two examples, the behavior of films could be hardly envisaged from that of powders. For Y(TFA)$_3$, no dependence on the wet/dry condition was observed and, for Ba(TFA)$_2$, the decomposition of films in air could not be deduced by progressively diminishing the initial mass of the powders (the dashed curve of Fig. 2 indicates that the decomposition temperature increases when the initial mass is reduced). When the reactive gas transport is the rate-limiting step, films always decompose at a temperature lower than powders. This general behavior should be taken into account in the present search for low-temperature processing of CSD oxide films. In the same line, we have recently shown [6] that CeO$_2$ films can be obtained at temperatures as low as 160°C even if precursor powders decompose at 300°C [7].

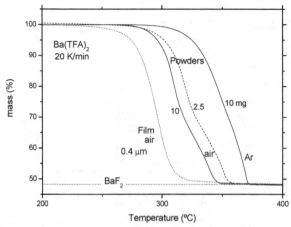

Figure 2.- TG curves of the decomposition of powders and films of Ba(TFA)$_2$ in air.

Heat transport

During an exothermic process, the sample temperature tends to be higher than its surroundings, this overheating being more pronounced for larger samples. When this effect has a moderate intensity, the TG curves tend to shift to lower temperature as the sample mass increases. This is the case of the decomposition of Ba(TFA)$_2$ powders in air (Fig. 2). The curve corresponding to 10 mg is shifted by 10°C with respect to that of 2.5 mg.

When the overheating is very pronounced (large powder masses or highly exothermic reactions) it may lead to thermal runaway process: the reaction rate increases due to the overheating and, consequently, more heat evolves per unit time what further increases the sample temperature and reaction rate. The local temperature may be hundreds of Celsius above the furnace temperature and, consequently, the sample decomposes very quickly. An abrupt mass loss step in the TG curve is thus a characteristic feature of this runaway (or combustion) process (TG curve of Y(TFA)$_3$ powders in air of Fig. 1). In addition, a sharp exothermic DTA peak is measured (Fig. 1) (the broader shoulder appearing on its high-temperature side is an artifact of the deconvolution procedure used to correct for the apparatus time constant).

For films, the substrate offers a very efficient dissipation path for the heat evolved during decomposition and, consequently, thermal runaway is very unlikely to occur. In fact, we have never observed the characteristic TG/DTA features related to thermal runaway in any film obtained from precursors that experience combustion when they are in the form of powders. In Fig 1 the TG curves of films are very smooth (the DTA signal was too low to be detected) in contrast with those of powders.

Combustion of thin films has been recently proposed as a new route to synthesize functional oxides on plastic substrates [3]. The fact is that the thermal analyses of ref. [3] were carried out on powders and not on films. We thus consider that our results put under suspicion the interpretation that combustion occurred in those films.

Out diffusion of volatiles and the stability of intermediate products

   A number of metal organic precursors tend to decompose into carbonates instead of oxides [1]. This is the case of precursors containing Ba. $BaCO_3$ is unstable above 1000 °C where it decomposes into its oxide. This fact has a negative effect on the fabrication of $YBa_2Cu_3O_{7-x}$ superconductor films because of the high amount of carbon present at the temperature of $YBa_2Cu_3O_{7-x}$ formation (around 800°C), that is then incorporated to the film decreasing the superconducting properties. This fact has driven the search for precursors containing fluorine; however, those may cause environmental problems [8].
   Probably, $BaCO_3$ is the result of a reaction between an intermediate Ba-containing species and a gaseous species (e.g. CO or CO_2) that evolves during decomposition (see the MS curves of Fig. 3a). If true, carbonation would be lower in films because the gaseous species would escape more easily from the sample. Fig. 3b confirms this prediction for the particular case of Ba propionate [$Ba(CH_2CH_2COO)_2$] decomposition in air.

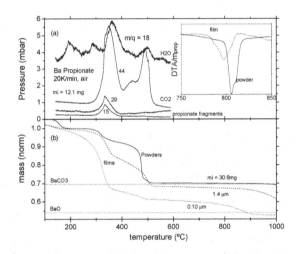

Figure 3.- Decomposition of Ba propionate in air: mass spectroscopy curves (a) and TG curves (b). Inset: DTA signal due to $BaCO_3$ allotropic transformation.

   The TG curves of Fig. 3b indicate that, after decomposition one obtains 100% of carbonate for powders and a lower amount for films. For a better determination of the carbonate content we have used the area of the $BaCO_3$ allotropic transformation peak appearing in the DTA curves around 800°C. The inset of Fig. 3 makes clear that carbonation is lower for the 1.4 μm thick film (the area normalized to the propionate mass is 30% lower for the film). For thinner films, Fig.3b seems to indicate that carbonation would be still lower; however we have not been able to measure the DTA signal in this case. Another interesting feature of thin films is that Ba carbonate decomposes into BaO at a temperature much lower than that of powders.

As far as we know, this is the first experimental evidence that carbonation of Ba can be reduced with metal organic precursors without fluorine. The significance of this result for the production of $YBa_2Cu_3O_{7-x}$ has to be evaluated by analyzing its decomposition of the mixture containing the precursors of the other metal oxides.

CONCLUSIONS

The examples discussed so far clearly show that films may decompose differently than powders. The differences are not only quantitative (e.g. films tend to decompose at a lower temperature) but also qualitative (e.g. films do not experience combustion). These results indicate that the thermal analysis of powders should be treated with caution when applied to films.

ACKNOWLEDGMENTS

We acknowledge the financial support from MICINN (MAT2011-28874-C02-01 and -02, Consolider Project NANOSELECT: CSD2007-00041) and by the Generalitat de Catalunya (Pla de Recerca 2009SGR-185, 2009-SGR-770 and XaRMAE).

REFERENCES

1. R.W. Schwartz, T. Schneller, R. Waser, Chemical solution deposition of electronic oxide films, C. R. Chim. **7**, 433 (2004).
2. J. Farjas, A. Pinyol, Ch. Rath, P. Roura, and E. Bertran, *Phys. Stat. Sol. A* **203**, 1307 (2006).
3. M-G. Kim, M.G Kanatzidis, A. Facchetti and T.J. Marks, Nature Materials **10**, 382 (2011).
4. P. Roura and J. Farjas, Thermochim. Acta **430**, 115 (2005).
5. H. Eloussifi, J. Farjas, P. Roura, J. Camps, M. Dammak, S. Ricart, T. Puig and X. Obradors, J. Therm. Anal. Calorim. DOI 10.1007/s10973-011-1899-5.
6. P. Roura, J. Farjas, S. Ricart, M. Aklalouch, R.Guzman, J. Arbiol, T. Puig, A. Calleja, O. Peña-Rodríguez, M. Garriga, and X. Obradors, Thin Solid Films **520**, 1949 (2012).
7. P. Roura, J. Farjas, J. Camps, S. Ricart, J. Arbiol, T. Puig, and X. Obradors, J. Nanopart. Res. **13**, 4085 (2011).
8. X. Obradors, T. Puig, A. Pomar, F. Sandiumenge, N. Mestres, M. Coll, A. Cavallaro, N. Roma, J. Gazquez, J.C. Gonzalez, O. Castano, J. Gutierrez, A. Palau, K. Zalamova, S. Morlens, A. Hassini, M. Gibert, S. Ricart, J.M. Moreto, S. Pinol, D. Isfort and J. Bock, Supercond. Sci. Technol. **19**, S13 (2006).

Mater. Res. Soc. Symp. Proc. Vol. 1449 © 2012 Materials Research Society
DOI: 10.1557/opl.2012.1040

Synthesis and magnetic properties of manganite thin films on Si by polymer assisted (PAD) and pulsed laser deposition (PLD).

J. M. Vila-Fungueiriño, B. Rivas-Murias, F. Rivadulla*

Centro de Investigación en Química Biológica y Materiales Moleculares (CIQUS),
C/Jenaro de la Fuente s/n, Campus Vida, Universidad de Santiago de Compostela,
15782-Santiago de Compostela, Spain.

ABSTRACT
We report the synthesis of polycrystalline films of $La_{1-x}Ca_xMnO_3$ in Si (111) by Polymer Assisted Deposition (PAD). An aqueous solution polyethyleneimine (PEI) and different metal ions stabilized with EDTA, was spin coated on hydrophilized Si substrates and subsequently annealed under different atmospheres. Homogeneous, dense polycrystalline films are obtained at optimized conditions of 950 °C under flowing O_2. The morphology and magnetic properties of the samples are compared with films obtained by Pulsed Laser Deposition.

INTRODUCTION
Manganites of general formula $La_{1-x}(Ca,Sr)_xMnO_3$ (LCMO, LSMO) can be tuned to present a high Curie temperature, T_C, and a large degree of spin-polarization [1,2] that makes them attractive for a large number of applications. For example, IR detectors or any other technology based on sensing a change in resistance in response to a temperature or magnetic field variation may be designed.[3] Also, they are being actively studied as polarized ferromagnetic (FM) electrodes for spin-injection in semiconductors. Particularly interesting in this regard is the direct deposition on Si, due to the long spin-coherence time of this semiconductor,[4] as well as to possibility of integrating the multifunctional oxide with this semiconductor. Different authors grew epitaxial thin films of LCMO and LSMO on Si through a buffer layer, by Pulsed Laser Depostion (PLD) and sputtering.[5] The nature of the buffer layer changes depending on the composition of the FM, and in some cases complex multilayer structures must be used between the FM and the semiconductor,[6,7] affecting the transport of charge carriers across the interface into Si. On the other hand, attempts to grow the FM oxide directly on Si are scarce. On the other hand, polycrystalline films of LSMO were deposited on Si by sputtering,[4] with magnetic properties similar to bulk manganites.
Given the interest in de deposition of multifunctional, multicationic oxides directly on Si, we have considered the possibility of chemical solution deposition (CSD) and spin-coating based methods, as they are cost-effective approaches which could be competitive with sputtering and PLD in this case. This is an approach that has been shown to be successful to grow epitaxial LCMO and LSMO on STO and LAO substrates,[8,9,10] but it has not been extended to substrates with different structure, like Si.
Here we report a comparison of the morphologic and magnetic properties of LCMO thin-films grown directly on Si by PAD [11] and PLD. The films deposited by PAD were subsequently annealed at different temperatures under different atmospheres to control their Curie temperature, T_C, and magnetic moment. The properties of the polycrystalline layers prepared by PAD are perfectly comparable to films grow by PLD at high temperature.

EXPERIMENT

Nitrates of the metallic cations were dissolved in Mili-Q water, to a final volume of 25 ml, to give final concentrations between 50-150 mM. An stoichiometric amount of EDTA was then added to each solution to form $[EDTA-M^{n+}]^{(4-n)-}$ complexes. An amount of polyethyleneimine, PEI, (PEI:EDTA=1:1 w/w) is then added to the solution, and stirred continuously at 50°C, until a transparent liquid is formed. At the acidic pH of the solutions most of the amines of PEI are protonated, acting as a cationic polymer that binds to $[EDTA-M^{n+}]^{(4-n)-}$. This results in a homogeneous distribution of the cations in the polymeric solution, at the time that provides an easy way to control viscosity and makes the solution stable for months. In order to remove non-coordinated cations and counter anions, solutions are filtered using 10 kDa Amicon filtration units in a ultracentrifuge. These solutions are analyzed by ICP to know the exact amount of cations retained. Typically, the degree of retention is higher than 75%, except in Mn, which is of the order of 50%. The solutions are then mixed in stoichiometric proportions according to the desired final composition. They are further concentrated by heating gently in a hot plate, in order to reach the optimum viscosity for spin-coating deposition.

Si(111) substrates were hydrophilized by immersed then in acetone for 5 minutes in an ultrasonic bath, before baking them for 15 min at 47°C in a bath of $H_2O:H_2O_2:NH_4OH$ (7:1:4).[13]

The solution is then spin-coated on hydrophilic Si(111) substrates at 5000 rpm, 20 s, rendering homogeneous polymeric films on the whole surface of the substrates. The films are then annealed at different temperatures and atmospheres, depending on the final composition. Thermogravimetric analysis (TGA) shows that the polymer is completely eliminated above 450°C in oxidizing atmospheres (air or pure O_2).

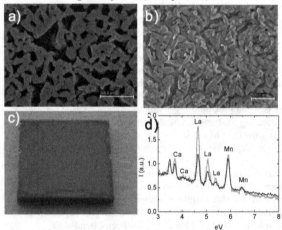

Figure 1: Effect of the annealing atmosphere on the morphology of the films (final temperature 950°C, 2 hours). a) corresponds to air, and b) to oxygen. The picture in d) shows the homogeneity of the oxygen film in a Si (111), 5×5 mm^2 substrate. d) Part of the EDX analysis of a $La_{2/3}Ca_{1/3}MnO_3$ film deposited by PAD (solid line) and PLD at 960°C (open circles, see text).

The final annealing temperatures were varied between 600°C to 1100°C, finding 950°C, 2h, as the optimum annealing conditions. We have found that the morphology of the film is highly dependent on the atmosphere (see figure 1). All the results discussed in this paper correspond to samples annealed in oxygen.

DISCUSSION

The films annealed in oxygen are composed of well sintered particles of LCMO, with a size ranging between ≈80 and 200 nm. In Figure 1d) we show an example of an EDX analysis performed in a $La_{2/3}Ca_{1/3}MnO_3$ film deposited on Si. The experimental cationic ratio from the EDX analysis, La:Ca:Mn=0.69(2):0.32(1):1 compares very well with the nominal composition 0.67:0.33:1. This shows that the PEI-EDTA method is able to reproduce the stoichiometry of the initial solutions into the film.

Regarding their magnetic properties all samples show a behavior similar to bulk, as we will discuss. The thickness of the films and hence the amount of magnetic material deposited can be controlled by tuning the total cationic concentration of the polymeric solution that is spin coated on the Si substrate. In figure 2 we show the hysteresis loops of equal composition LCMO films, as a function of the concentration of the initial solution. The magnetic response scales very well with the concentration.

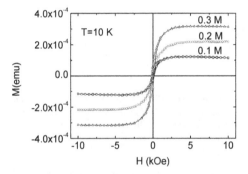

Figure 2: Hysteresis loops of LCMO films on Si (111), from starting solutions with different ionic concentration.

This is an important advantage of this method with respect to physical deposition techniques, like sputtering or PLD, that require an increasing time of deposition to increase the thickness of the films.

One of the most interesting conclusions of this work is that anisotropic, bar-shaped LCMO structures can be grown on Si, just by controlling the initial concentration of ions in the polymeric solutions. We have found, that for total ionic concentrations lower than ≈0.2 M, the film is composed by a dense matrix of elongated particles (see Figure 3). Sizes range between 100-400 nm wide and 0.8 to 4 μm long.

Recently, synthesis of bar-shaped LSMO films on Si was reported by a combined template-assisted plus CSD approach.[13] The aspect ratio and magnetic properties are very similar to these reported here for LCMO. So, our study shows that there are other alternative, most simple techniques, for the preparation of anisotropic LCMO on Si substrates. Also, we have tried to grow the same type of structures on LAO and STO, using low ionic concentrations, but the attempts were unsuccessful, showing that having a dissimilar structure between the film and the substrate is important to grow anisotropic LCMO.

As previously reported, the magnetic response of these films are anisotropic (see Figure 3), with an easy plane magnetization attributed to the bars lying on the plane of the substrates. Surprisingly, we have found a similar in-plane/out-of-plane anisotropic magnetization in all the samples of this study, irrespective of the shape of the particles.

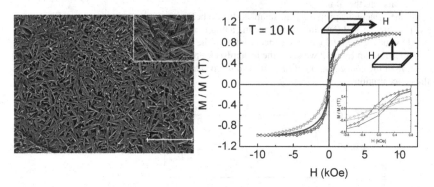

Figure 3: Left: Anisotropic, bar-like structures of LCMO @ Si. The inset of the upper corner shows a magnification of one area of the picture. Right: Hysteresis loops of bar-shaped LCMO, with the applied field parallel and perpendicular to the film plane. The blue solid line represents the hysteresis loop for the perpendicular orientation, after correction for the demagnetizing field effect.

In order to get an insight of the origin of this effect, we have measured the electron spin resonance of various films, above their Curie temperature. In Figure 4, it can be seen that the resonance field in the perpendicular orientation is always larger than in the parallel configuration, even above T_C, where no magnetic order occurs. This rules out the explanation based on a strong shape anisotropy due to the elongated bars, and points to a demagnetizing field effect as its most plausible origin. For instance, for a uniformly magnetized (infinite) plane the demagnetizing factor is 4π normal to the plane, and zero along the plane. So, the internal magnetic field seen by the LCMO particles must be recalculated, according to $H_{int}=H_{appl}-NM$, where H_{int}, H_{appl}, N, and M are the internal field, applied field, demagnetizing factor and the magnetization, respectively. The actual amount of magnetic material in the film was obtained from comparison of the intensity of the ESR data with an standard. This is a non-destructive, very sensitive technique, able to determine the total amount of magnetic material in the film, so comparison of the absolute magnetic response of different films can be made.

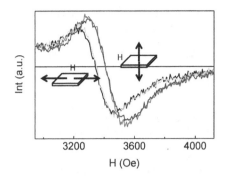

Figure 4: Electron Spin Resonance of a LCMO@Si film at 298 K, well above its T_C.

The perpendicular M vs. H loop, corrected for the demagnetizing field effect, is shown as a solid line in Figure 3. There is now a good overlap for both configurations, parallel and perpendicular, after the actual internal field is considered.

Finally, we have made a comparison between the films deposited by PAD and PLD. The main results are summarized in Figure 5.

The films deposited by PLD at room temperature are amorphous, and a subsequent annealing at 950°C in O_2 results in agglomeration into micron-sized particles.

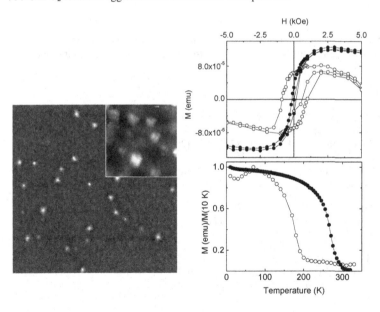

Figure 5: Left: Morphology of PLD films deposited at 950°C, $P(O_2)$=100 mTorr. The inset shows a film growth at room temperature, after an annealing at 950°C in O_2(inset). The size of the picture is 10×10 μm^2 in both cases. Right top: Magnetic moment vs field at 10 K for LCMO@Si films grown by PLD at 960°C (open symbols) and PAD (closed symbols). Right bottom: Temperature dependence of the magnetic moment normalized by its value at 10 K, for the same films.

Much smoother films are grown when the substrate is held at 950°C under an oxygen pressure of 100 mTorr. However, a comparison of the magnetic data with those of PAD for the a priori same composition, shows evident differences: the T_C is greatly reduced and the coercive field is larger than in the corresponding PAD film. This is consistent with a substantial imbalance in the La/Ca ratio in the film, which does not retain the composition of the target. This was confirmed by EDX (see Figure 1) which shows that the film is La rich, and hence does not retain the concentration of holes of the target. To keep the stoichiometry will require a careful optimization of the temperature, oxygen pressure, laser energy and frequency.

CONCLUSIONS

We have shown that dense polycrystalline films of LCMO can be deposited in Si by PAD. This method allows a good control of the stoichiometry and thickness. Moreover, the shape of the particles in the films is highly dependent on the initial total ionic concentration in the polymeric solution. We compared these results of films grown by PAD with those grown by PLD. The comparison shows that PAD offers many advantages over PLD for the deposition of polycrystalline LCMO films on Si.

ACKNOWLEDGMENTS

We thank financial support from Ministerio de Economía y Competitividad (MINECO, Spain) through the project MAT2010-16157. J.M. V-F also acknowledges MINECO for support with a PhD grant (FPI program).

REFERENCES

[1] P. Schiffer, A. P. Ramirez, W. Bao, S-W. Cheong, Phys. Rev. Lett. **75**, 3336 (1995).
[2] H. Y. Hwang, S-W. Cheong, N. P. Ong, B. Batlogg, Phys. Rev. Lett. **77**, 2041 (1996).
[3] J.-H. Kim, A. M. Grishin, Appl. Phys. Lett. **87**, 033502 (2005).
[4] I. Bergenti et al. J. Mag. Mag. Mater. **312**, 453 (2007).
[5] A. K. Pradham et al., Appl. Phys. Lett. **86**, 012503 (2005).
[6] D. Kumar et al., Appl. Phys. Lett. **78**, 1098 (2001).
[7] J.-H. Kim, S. I. Khartsev, A. M. Grishin, Appl. Phys. Lett. **82**, 4295 (2003).
[8] M. Jain et al. Appl. Phys. Lett. **88**, 232510 (2006).
[9] R. Cobas et al. Appl. Phys. Lett. **99**, 083113 (2011).
[10] L. Fei et al. Appl. Phys. Lett. **100**, 082403 (2012).
[11] Q. Jia, T. M. McCleskey, A. K. Burrell, Y. Lin, G. E. Collins, H. Wang, D. Q. Li, S. R. Foltyn, Nat. Mat. **3**, 529 (2004).
[12] V. E. Agabekov et al., Russ. J. Gen. Chem. **77**, 343 (2007).
[13] A. Carretero-Genevrier et al. Adv. Mater. **20**, 3672 (2008).

Mater. Res. Soc. Symp. Proc. Vol. 1449 © 2012 Materials Research Society
DOI: 10.1557/opl.2012.812

Ink-jet Printing of YBa2Cu3O7 Superconducting Coatings and Patterns from Aqueous Solutions

Isabel Van Driessche[1], Jonas Feys[1], Pieter Vermeir[1], Petra Lommens[1]

[1]SCRiPTS, Department of Inorganic and Physical Chemistry, Ghent University, Krijgslaan 281 – S3, 9000 Gent, Belgium, Isabel.vandriessche@ugent.be.

ABSTRACT

In this paper, we combine the use of Drop-on-Demand (DOD) ink-jet printing with completely water- based inks as a novel approach to the CSD process for coated conductors. This method holds the promise of improved scalability due to lower ink losses, continuous processing and a drastically increased precursor lifetime due to the prevention of solvent evaporation and dust incorporation. Moreover, ink-jet printing has the potential to switch quite easily from continuous coatings to a multi-filamentary pattern, which is particularly important for alternating current (AC) or field applications of coated conductors. The fluid properties, often expressed with dimensionless constants, like the Reynolds and Weber numbers, for printable liquids were determined. For proof-of-concept, single crystals of $SrTiO_3$ with a low mismatch towards YBCO, were used as substrates.

INTRODUCTION

The production of low cost, long length $YBa_2Cu_3O_{7-\delta}$ (YBCO) coated conductors is one of the main prerequisites for spreading the use of superconductivity in power applications. Currently, a promising coated conductor design is based on a metallic Ni-5%W tape (RABiTS) coated with a $La_2Zr_2O_7 – CeO_2$ buffer structure, and a superconducting YBCO top layer [1]. Up to now, vacuum techniques have resulted in the best properties for high temperature superconductor (HTSC) thin films. However, to reduce production costs and improve scalability, a shift towards chemical solution deposition (CSD) conditions would be preferred. The main advantages are the lower investment, faster deposition with higher yield, ease of stoichiometric composition control and modification, and processing under ambient pressure, enabling completely continuous production [2].

Inkjet printing is widely used for the fabrication of coatings and patterns onto a variety of substrates . It is a simple and cost-effective technique for ceramic coatings. The reproducible dispensing of ink droplets in the range of pL to nL volumes at high rates (kHz) allows for high 3D resolution, strict control of the thickness and gradient porosity. Inkjet systems can be readily scaled-up for industrial manufacturing and the technology is environmentally friendly utilizing only the exact amount of necessary material.

In the present manuscript, drop-on-demand ink-jet printing of coatings and tracks will be reported using piezoelectric printing systems, on both single crystals and industrially-relevant metal substrates, using inks formulated based on non-fluorine chemistries.

EXPERIMENT

Water-based precursor solutions were prepared by dissolving stable, cost-effective and easily available inorganic salts in an aqueous solution of coordinating ligands. As a result, solvation by water molecules is discouraged and neither extensive hydrolysis nor precipitation is likely to take place, resulting in very homogeneous materials [3]. The aqueous precursor was prepared starting from $Y_2(CO_3)_3.1.9H_2O$ (99.9 %, Sigma Aldrich), $Ba(OH)_2.8H_2O$ (98 %, Janssen) and $Cu(NO_3)_2.2.5H_2O$ (98 %, Alfa Aesar) salts dissolved in water and nitrilo-triacetic acid (NTA, 99 %, Alfa Aesar) in a 0.45 : 1 ratio for NTA : total metal concentration. The addition of triethanolamine (99+ %, Acros Organics) increases the pH and the viscosity to the desired values of 6-8 and 4.77 mPa s (25 °C, 100 rpm, Brookfield DV-E Viscometer) respectively. Attention was given towards the development of an ink with neutral pH and non-aggressive components to prevent corrosion effects inside the printing system. The total metal concentration of the precursor solution was 1.1 mol/L (0.185 mol/L YBCO), as verified by ICP-OES (Spectro, Genesis). By slow evaporation of the solvent (water) at 60°C, condensation of the complexes in the solution takes place, leading to the formation of a homogeneous gel network. A stable shelf-life of several months was established. It has been reported previously that this precursor system can be used for several ceramic coatings [4-8], buffer layers [9-10] and superconductors [11]. The samples were heat treated at 20 °C/min to 780 °C for 2 h in a humid O_2/N_2 atmosphere, followed by an oxygenation step at 520 °C, with a dwell at 400 °C.

Laboratory-scale printing was performed using piezoelectric micro-dispensers from Microfab Technologies, Inc. (USA) mounted on X-Y positioning stages under computer control in a clean room environment. Printing was performed by moving the nozzle over the substrate at constant velocity and jetting at a constant frequency (typically 1 kHz), the ratio of these controlling the inter-droplet spacing (typically 75 μm).

To characterize jetting behaviour and allow the optimisation of ink and printing parameters, an optical drop visualisation system was developed comprising a high-sensitivity camera with 1292×964 px resolution at 30 frames/s (Allied Vision Technologies, Stingray F-125B) and a telecentric zoom lens (ML-Z07545, Moritex). Collimated, strobed LED illumination was used in a backlit configuration. The camera shutter and LED strobe were synchronized with droplet ejection with a selectable delay time, such that each frame corresponded to the state of the ink stream a chosen time after ejection.

The critical temperature of the superconducting layers was measured by resistivity measurements as a function of temperature using a custom-made four-point test device (Keithley). The critical current was determined from the third harmonic of the induced signal in a pick-up coil from an AC drive signal, using a Theva Cryoscan setup in liquid nitrogen. For this inductive measurement, a constant-voltage criterion of 50 μV was selected.

The composition, crystallinity and texture of the processed films was verified using X-ray diffraction, both in the Bragg-Brentano configuration for phase identification and configured for texture analysis (Thermo Scientific ARL X'TRA and Siemens D5000; Cu-K_α). The sample morphology was characterized using optical microscopy (Leitz, Laborlux 12 POL S) and SEM (FEI Quanta 200 FEG). A cross section of the layers was made using a FIB module coupled with

SEM to verify the thickness of the layers. The topology of multi-filamentary patterns was visualized using optical and AFM profilometry (Veeco NT9080).

DISCUSSION

The generation of droplets in a DOD printer is a complex process, and the precise physics and fluid mechanics of the process are the subject of much research [12]. The behaviour of inks in the printing system can be quantified by a number of dimensionless groupings of physical constants, i.e. the Reynolds (Re), Weber (We) and Ohnesorge (Oh) numbers:

$$Re = \frac{\upsilon r \rho}{\eta}, We = \frac{\upsilon^2 r \rho}{\sigma}, Oh = \frac{\sqrt{We}}{Re} = \frac{\eta}{\sqrt{\sigma \rho r}}$$

with σ, ρ, η and υ the ink surface tension (J m^{-2}), density (kg m^{-3}), viscosity (Pa s) and velocity (m s^{-1}) respectively and r the radius of the orifice of the nozzle (m). The Reynolds number is a ratio of internal and viscous forces and the Weber number shows the ratio between internal and surface tension forces. The inverse value of the Ohnesorge number is a characteristic dimensionless number which is independent of droplet velocity. Often it is written that Oh^{-1} should be between 1 and 10 for proper jetting properties. If the ratio is too low, viscous forces become more dominant preventing drop ejection; conversely, if the ratio is too high the possibility for satellite droplet formation becomes high [13-15]. In Table 1, we show the different numbers calculated for our ink, taking into account the ink viscosity obtained at the highest shear rate. For Oh^{-1}, a value of 7.37 is calculated, which is within the desired range.

Table I. Fluid properties of the YBCO ink for an orifice radius (r) of 3 x 10^{-5} m.

Type of ink	Surface tension σ [J m^{-2}]	Density ρ [kg m^{-3}]	Viscosity η [Pa s]	Orifice diameter (μm)	Re	We	Oh^{-1}
Water-based	6.79×10^{-2}	1233	6.8×10^{-3}	30	15.77	4.58	7.37

From Figure 1, one can see that initially, the drops form a liquid column which transforms into the actual droplet and an elongated tail. Here, breaking up of this tail from the droplet leads to the formation of a satellite drop. The presence of these satellite droplets should be avoided at impact with the substrate, as the key goal is to leave a single isolated droplet to optimize precision, resolution and accuracy during printing. Therefore, the distance between the nozzle and the substrate should be chosen in such a way that the two droplets can merge before impact. On the other hand, an increased standoff will reduce the accuracy, because drag from air currents in the printing chamber makes the droplets deviate from their vertical trajectory, so the distance should be set as low as possible. In our case, the optimal distance between print head and substrate was determined to be between 0.6 mm and 2 mm.

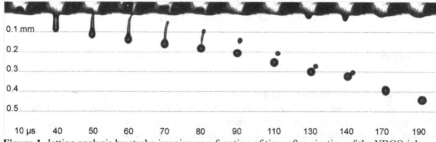

0.1 mm

0.2

0.3

0.4

0.5

10 µs   40   50   60   70   80   90   110   130   140   170   190

Figure 1. Jetting analysis by strobe imaging as a function of time after ejection of the YBCO ink from the nozzle.

The estimated volume and velocity of droplets are within the range of 65 to 80 pl and 2.3 to 3.5 m/s respectively within the printing parameters tested. For the experiment displayed in Figure 1, the volume is around 75 pl with a droplet velocity of 2.9 m/s after merging at 190 µs. At 110 µs, the estimated volumes for the main droplet and the satellite are close to 60 pl and 15 pl respectively.

In figure 2, optical and profilometry images are presented of printed wet droplets, dried at 100°C. The diameters and heights of the droplets in the two left images vary in the ranges 145 - 160 µm and 40 - 160 nm respectively and, for the pattern, between 120 - 140 µm and 170 - 200 nm respectively. Due to the small volume of material in one droplet or a single printed line, with rapid solvent evaporation from the edges, a clear "coffee-ring" effect can be observed in the profilometry measurements.

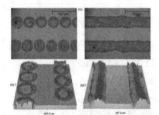

Figure 2. Optical micrographs and interference profilometry profiles for droplets of the YBCO water-based ink, printed with longitudinal spacings of (a) 0.2 mm, (b) 0.15 mm and (c) 0.1 mm, demonstrating that both continuous coating and patterning are possible.

When bringing the droplets closer together in the axial direction, it becomes equally possible to change from printing a pattern to achieving complete coverage of the substrate.The thickness after complete heat treatment of the YBCO thin films varies between 310 and 400 nm. Microstructural investigation of the top layer by SEM, is showing a crack-free and dense surface. Though, secondary phases enriched by Ba and Cu, are visible. The morphology will be further improved by an optimisation of the heat treatment.

XRD patterns reveal the formation of crystalline films of pure YBCO (figure 3). The in-plane and out-of-plane misorientation of the YBCO film was further characterized by a φ-scan (figure 3b) and ω-scan (figure 3a, inset). An average full width at half maximum (FWHM) value of 1.87° for the (103) φ-scan and a FWHM of 0.68° for the (005) ω-scan proves that highly textured YBCO was obtained.

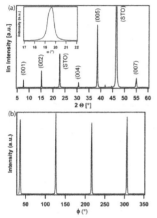

Figure 3. (a) θ-2θ scan and ω-scan of the (005) plane (inset) and (b) φ-scan obtained for the (103) plane of the water-based YBCO ink after complete conversion with an average FWHM for the last two of 0.68° and 1.87° respectively.

The result evidences a high critical temperature $T_{c(50)} \geq 92$ K and a sharp transition into the superconducting state. A Jc of 0.7 MA/cm^2 was determined.

CONCLUSIONS

We have successfully deposited complete coatings and multi-filamentary structures of YBCO on a single crystal SrTiO$_3$ substrate using a DOD piezoelectric ink-jet printing system. A new and stable YBCO ink was prepared starting from low cost and environmentally benign metal salts. This ink was characterized for its jetting properties as well as for the wetting on the substrate. With a ratio Re/We$^{1/2}$ of 7.37 and a surface tension of the liquid within the five degree wetting envelope for a cleaned SrTiO$_3$ substrate, the ink rheology is well within the criteria for good jetting and wetting behaviour. After optimization, printing of multi-filamentary YBCO patterns became possible by carefully tuning the printing parameters. Tracks 200 nm thick and 200 μm wide, were deposited and heat treated. The tracks have sharp edges and are very homogeneous in their overall profile, showing almost no coffee ring effect. A full YBCO coating with a final thickness of 310 – 400 nm and exhibiting strong c-axis orientation was also obtained, with a critical current density of 0.67 MA/cm^2 (I_c = 23.5 A/cm) at 77 K in self-field.

ACKNOWLEDGMENTS

This research was funded by the European Union under the FP7 Frame work : EU project EFECTS (FP7-NMP-2007-SMALL-1 grant n°205854).

REFERENCES

1. K. Knoth, S. Engel, C. Apetrii, M. Falter, B. Schlobach, R. Hühne, S. Oswald, L. Schultz, B. Holzapfel, Current Opinion in Solid State and Materials Science 2006, 10, 205.
2. Van Driessche I, Penneman G, Bruneel E, Hoste S, Pure Appl Chem, 2002, **74**, 2101-9.
3. Van Driessche I, Penneman G, Abell JS, Bruneel E, Hoste S, Thermec'2003, Pts 1-5, 2003, **426-432**, 3517-22.
4. De Buysser K, Lommens P, de Meyer C, Bruneel E, Hoste S, Van Driessche I, 2004, **48**, 139-44.
5. Gryglewicz G, Stolarski M, Gryglewicz S, Klijanienko A, Piechocki W, Hoste S, Van Driessche I, Carleer R, Yperman J, 2006, **62**, 135-41.
6. Vergote GJ, Vervaet C, Van Driessche I, Hoste S, De Smedt S, Demeester J, Jain RA, Ruddy S, Remon JP, Int J Pharm, 2002, **240**, 79-84.
7. Le MT, Van Well WJM, Van Driessche I, Hoste S, Applied Catal A, 2004, **267**, 227-34.
8. M. Arin, P. Lommens, N. Avci, S. C. Hopkins, K. De Buysser, I. M. Arabatzis, I. Fasaki, D. Poelman and I. Van Driessche, *J. Eur. Ceram. Soc.* **31**, 1067-1074 (2011).
9. Penneman G, Van Driessche I, Bruneel E, Hoste S, Euro Ceramics Viii, Pts 1-3, 2004, **264-268**, 501-4.
10. Van Driessche I, Penneman G, De Meyer C, Stambolova I, Bruneel E, Hoste S, Ttp, Euro Ceramics Viii, Pts 1-3, 2002, **206-2**, 479-82.
11. J. Feys, P. Vermeir, P. Lommens, S. C. Hopkins, X. Granados, B. A. Glowacki, M. Baecker, E. Reich, S. Ricard, B. Holzapfel, P. Van der Voort and I. Van Driessche, *J. Mater. Chem.* **22**, 3717-3726 (2012).
12. B. Derby, in *Annual Review Of Materials Research*, Vol. 40, Annual Reviews, Palo Alto 2010, 395.
13. B. Derby, N. Reis, K. A. M. Seerden, P. S. Grant, J. R. G. Evans, Solid Freeform and Additive Fabrication-2000 2000, 625, 195.
14. R. Noguera, M. Lejeune, T. Chartier, J Eur Ceram Soc 2005, 25, 2055
15. J. Windle, B. Derby, J Mater Sci Lett 1999, 18, 87.

Mater. Res. Soc. Symp. Proc. Vol. 1449 © 2012 Materials Research Society
DOI: 10.1557/opl.2012.1250

Growth of epitaxial CeO₂ buffer layers by polymer assisted deposition

A. Calleja[1], R. B. Mos[1,2], P. Roura[3], J. Farjas[3], J. Arbiol[1,4], L. Ciontea[2], X. Obradors[1], T. Puig[1]

[1] Institut de Ciència de Materials de Barcelona-Consejo Superior de Investigaciones Científicas (ICMAB-CSIC), Bellaterra, Catalonia, Spain

[2] Technical University of Cluj, Cluj-Napoca, Romania

[3] GRMT, Dept. of Physics, University of Girona, Campus Montilivi, Edif. PII, E17071 Girona, Catalonia, Spain

[4] Institució Catalana de Recerca i Estudis Avançats (ICREA), 08010, Barcelona, Catalonia, Spain

ABSTRACT

Polymer assisted deposition (PAD) has been reported as a novel CSD approach for thin film growth with improved homogeneity and long stability by forming a metal polymer species. It also offers the interesting possibility of having a library of PAD solutions for each precursor metal and obtaining the required composition by simple mixing. Another potential advantage is the increase in thickness since mechanical stresses are expected to be alleviated during shrinkage in the metalorganic decomposition by the metal-polymer network.
Cerium oxide films on YSZ single crystals were grown from water-based solutions containing cerium nitrate, polyethyleneimine and complexing EDTA, in order to explore the benefits of using the PAD approach for growing buffer layers in coated conductors. An ultrafiltration step was performed to remove the non-coordinated species in solution. The degree of purification and efficiency in the cerium recovery was investigated by different techniques. TGA-DTA analysis was used to provide guidance to the best thermal profiles in different atmospheres and specially to diminish the adverse effects of exothermic events during decomposition. Microstructural evolution was tracked by AFM and TEM, while epitaxial fraction was followed by X-ray diffraction. The results show the high importance of choosing the proper atmosphere and the need for tuning of heating ramps to obtain dense, flat and epitaxial ceria films by PAD.

INTRODUCTION

Ceria (CeO₂) is an important component in many technological devices due to its remarkable properties [1]. Most of the technological applications envisage its use in catalysis [2], electrochemistry [3] and optics [4]. Lately, the cerium oxide (CeO₂) thin films have been widely used in the fabrication of coated conductors as a promising buffer layer due to the good chemical compatibility and lattice match with $YBa_2Cu_3O_7$ (YBCO) ($\varepsilon = -0.5$ %). There is a clear need for full understanding of surface chemistry and morphology of CeO_2 thin films on the properties of the final superconducting film.

Compared with the vacuum deposition methods, chemical solution deposition [5-7] requires less equipment and less complexity for upscaling. Recently, Q.X. Jia et al. [8] have reported a general methodology for the epitaxial growth of thin films on single crystalline substrates using inorganic salts as precursors obtained by a new chemical method called Polymer Assisted Deposition (PAD). In the PAD process, an aqueous solution of metal precursors is mixed with a soluble polymer to obtain a metal-polymer complex by ultrafiltration. The major distinction between the PAD process and other chemical solution techniques lies in the soluble polymer that plays a significant role since it helps in achieving the desired viscosity, acting as a complexing agent to the metal salt without counterions, thus ensuring a homogeneous metal distribution and preventing unwanted reactivity that can lead to the formation of undesired phases.

The process begins with the binding of a water soluble metal salt (such as metal nitrate, metal chloride, or metal hydroxide) to a polymer, such as polyethyleneimine ($-CH_2-CH_2-NH-$)$_n$ (PEI). Some metal ions strongly bind to PEI, such as copper (II), while others, such as cerium (III), are preferably bound to functionalized PEI or to the ethylenediaminetetraacetic acid (EDTA), forming an EDTA-metal complex. The major advantage of the EDTA route is that EDTA forms stable complexes with almost all metals. The EDTA complexes further bind to the PEI via a combination of hydrogen bonding and electrostatic attraction [8-10]. At this respect, pH should be properly controlled to drive the chemical speciation in solution towards binding with the polymer.

In this work we report on the fabrication and characterization of cerium oxide thin films deposited on single crystalline (100)YSZ substrates ($\varepsilon = -4.6$%) starting from cerium (III) nitrate, EDTA and PEI dissolved in water. EDTA complexing agent was used to stabilize the cation, while PEI further enhances the complexing capacity due to the amine groups and ensures the solubility in water under almost neutral conditions.

EXPERIMENTAL

The coating solution was prepared by dissolving cerium (III) nitrate ($Ce(NO_3)_3$), ethylenediaminetetraacetic acid (EDTA), polyethyleneimine (PEI) with $M_w=70,000$ in water in 1:1:20.5 stoichiometric ratio. $EDTA^{4-}$ forms a strong 1:1 complex with Ce^{3+} [11]. The idea of using EDTA is the reduction of the concentration of free Ce^{3+} in the solution by the formation of soluble chelate complexes. The equilibrium constant for the reaction $Ce^{3+} + EDTA^{4-} \leftrightarrow [Ce(EDTA)]^-$ is $10^{15.98}$ [12]. The pH of the solution was adjusted with PEI to 5. After stirring, the solution was placed in an Amicon ultrafiltration unit containing a PM10 ultrafiltration membrane. The concentration of the precursor solution was adjusted to $[Ce^{3+}]=0.1$ M, respectively, as checked by using both a Varian Liberty 220 (Varian Inc., Palo Alto, CA) inductively coupled plasma – atomic emission spectrometer and a classical gravimetry with sulfuric acid and weighting the formed CeO_2. The viscosity of the solution measured at room temperature with a Thermo-Haake Rheostress 600 rheometer was 3.4 mPa·s. The long stability of the precursor solution (> 1 year) is highlighted.

The thermogravimetric-differential thermal analyses (TGA-DTA) were performed on the dried precursor gel from ambient temperature up to 650 °C in oxygen and nitrogen atmosphere, with a flow of 0.6 l/min and heating rates of 10 °C/min. The precursor solution and the reagents were characterized by FT-IR spectroscopy using a Perkin Elmer equipment.The precursor solution was spun on (001)YSZ substrates at a spinning rate of 6000 rpm for 2 min. After spin-coating, the samples were heated in a tubular furnace up to 1000 °C, then held for 8 h to allow completion of the epitaxial growth process, and finally cooled down to room temperature. The film growth was carried out both under inert (N₂) and oxygen (O₂) atmosphere as well as combinations of both gases.

These deposition and growth conditions yield films with a final thickness in the range of 10-40 nm, as determined from spectroscopic ellipsometry measurements. The influence of the thermal treatment on the structural properties and surface morphology of the films was investigated by X-ray diffraction both in θ-2θ and pole figure configurations and atomic force microscopy (AFM). The AFM images were analyzed using the commercial software package Mountains (Digital Surf). The percentage of flat area or planarity was calculated following the work of M. Coll et al. [13].

The microstructural characterization of the samples was carried out by means of transmission electron microscopy (TEM) . In order to obtain the high resolution TEM results we used a field emission gun microscope Jeol 2010F, which works at 200 kV and has a point to point resolution of 0.19 nm. Power spectra analysis were obtained from the HRTEM images.

RESULTS AND DISCUSSION

The thermal analysis of the precursor gel was performed separatedly in oxygen and nitrogen atmosphere in order to obtain information on the decomposition process of the deposited films and to establish a suitable heat treatment (Fig.1). One should take into account that nitrates are powerful oxidizers and together with the large excess of organic material, creates conditions where strong redox reactions are likely.

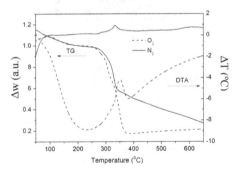

Figure 1. TGA-DTA curves for the decomposition of the cerium precursor in oxygen and nitrogen flow.

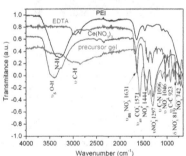

Figure 2. FT-IR spectra of the reagents and precursor gel.

In oxygen flow, two decomposition processes are observed at 307 and 345°C. At 370°C the decomposition ends, leaving a residue of about 18% of the initial dried mass. These processes correlate with two exothermic peaks (the first one detected as a shoulder) on the DTA curve.

In the case of nitrogen flow, the mass loss between 290 and 370 °C is only 40 wt.% of the initial mass of the dried precursor, while in O_2 reaches 80 wt.%. After this event, a progressive mass loss continues to take place up to 650 °C. In addition, the heat generated during the low temperature decomposition process is lower in nitrogen atmosphere. If we integrate the signal of the exothermic peak of the DTA curve up to 400 °C in N_2 atmosphere, we obtain 168 °C•s/mg, while under O_2 the value is 564 °C•s/mg. This is in agreement with the more exothermic reaction occuring in oxidizing atmosphere.

The IR active groups of the reagents and of the precursor gel were identified in the FT-IR spectra. As seen in Fig. 2, the vibration mode of the carboxyl group at 1684 cm^{-1} present in the FT-IR spectra of the pure EDTA is shifted towards higher wavelengths (1565 cm^{-1}) in the precursor solution [Ce(EDTA)PEI], indicating complexation with cerium [14]. Furthermore, the vibration modes of NO_3^- group have been also identified in the precursor gel, evidencing that some nitrate fraction is trapped in the retentate during ultrafiltration and therefore can further contribute to the exothermic signal observed in the DTA data.

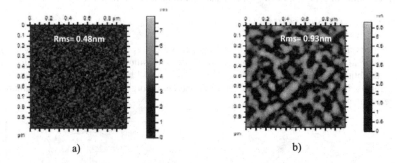

a)   b)

Fig. 3. AFM images corresponding to the PAD ceria films after thermal decomposition at 600°C in nitrogen (a) and 400°C in oxygen (b), showing the large differences in microstructure at this processing stage.

To study the effect of the composition of the gas flow, AFM was performed after finishing the organic decomposition according to the obtained TGA-DTA data (Fig. 3). Comparing both microstructures, an inhomogeneous, porous film with large voids distributed throughout the film is observed in the case of oxygen atmosphere, whereas a homogeneous and smooth film forms for the nitrogen flow. The high and fast exothermic signal registered in oxygen is an indication of a large release of decomposition gases through the bulk of the film, which is at the origin of the film roughness. Note that cerium oxide mobility at this temperature is very small, even at high temperature treatments. The film microstructure is not properly reconstructed and film flatness and percolation is not reached. This confirms that N_2 pyrolysis is necessary for ensuring high quality grown films.

Figure 4a shows typical (111) pole figures for a (100) CeO_2 film derived from the PAD precursor solution and for the (111)YSZ single-crystal substrate (Fig. 4b) after a growth in oxygen up to 1000 °C for 8 h. Single cube-on-cube heteroepitaxy is evident from this figure. The FWHM value of the (111) CeO_2 phi-scan was 2.5° (Fig. 5).

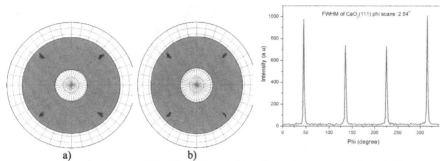

a)	b)

Figure 4. X-ray pole figures of (a) CeO₂(111) and (b) YSZ(111) substrate

Figure 5. φ scan of a CeO₂ buffer layer on YSZ substrate

AFM was used to characterize the surface morphology of the epitaxial ceria films grown on the (100) YSZsubstrates. Figure 6a and b display the AFM images and the corresponding planarity of the CeO_2/YSZ(100) sample annealed at 1000 °C in O_2 and in N_2 atmosphere, respectively.

The roughness values indicate that the ceria thin film annealed at 1000 °C in O_2 (Fig. 6a) is smoother and presents a higher flat area fraction of ~ 96 % with respect to the sample annealed in nitrogen (Fig.6b), which displays a rougher surface, showing larger and faceted grains. Therefore, this suggests that oxygen atmosphere enhances atomic mobility, enabling smooth and epitaxial CeO_2 layers.

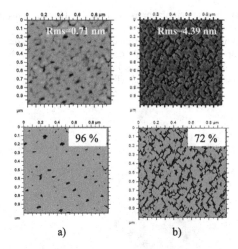

a)   b)

Figure 6. AFM images of different CeO₂ buffer layers grown on YSZ single crystal and their corresponding binary images obtained after applying a threshold mask of 1.5 nm [11]. The samples were treated under different conditions: (a) under oxygen and (b) under nitrogen. All the samples were heat treated at a heating rate of 10°C/min followed by a dwell of 8 h at 1000 °C. Values of rms roughness and flat area fraction for each sample are also indicated as legends.

Also mixed atmospheres were analyzed to check whether denser films could be obtained. In particular, one of such experiments is described here, consisting in a slow heating ramp of 2°C/min in nitrogen up to 600°C, followed by a 1-h dwell at this temperature. Then, the gas flow was switched to oxygen followed by a heating at 10°C/min up to 1000°C and a dwell of 8 h. In this case, XRD data shows that the film is polycrystalline since the (111) reflection of CeO₂ is clearly detected (Fig. 7a). Furthermore, the AFM analysis in Fig. 7b shows that surface roughness lies between the values obtained in pure oxygen and nitrogen. Grain distribution and size is qualitatively closer to the case of oxygen flow.

A high resolution TEM image near the interface with the substrate for this sample is shown in Fig. 8, where the grain boundaries of a few ceria particles are profiled with dashed white lines. It is to be noticed that misoriented grains with incoherent grain boundaries appear while interfacial grains are found, which grow with cube-on-cube epitaxial relationship with the YSZ substrate. The average height of the polycrystalline grains is approximately 30 nm, while for the epitaxial CeO₂ in the interface lies below 10 nm. The polycrystalline fraction is thought to arise   from

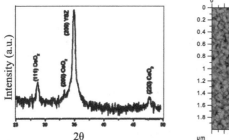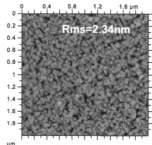

Fig.7. θ-2θ XRD pattern and AFM image corresponding to the PAD ceria film with modified heat treatment process.

unwished coarsening of the bulk nanocrystalline grains during the slow heating ramp and dwell at 600°C, which impedes grain boundary reconstruction to achieve full epitaxiality [15]. This is in contrast with the full epitaxial films obtained when a faster heating rate of 10°C/min is used, either in oxygen or nitrogen as gas flows. In any case, these results shows the importance of fine tuning of the processing parameters during heat treatment in order to obtain PAD films with suitable crystalline orientation, grain size and densification.

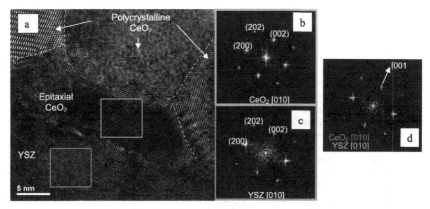

Figure 8. HRTEM image of the near surface region of the PAD CeO₂ film grown with modified heat treatment (a). SAD analysis of the squared regions (b) and (c), corresponding to the interfacial ceria grain and substrate, respectively. False color RGB image showing correct cube-on-cube epitaxy between CeO₂ and YSZ (d).

CONCLUSIONS

We have demonstrated the preparation of epitaxial CeO_2 thin films on YSZ single crystals from a novel water-based cerium solution, using PAD as a new chemical solution process. The large influence of the oxidizing power of the gas flow on the thermal behaviour of the precursor solution was confirms that exothermic events during organic decomposition are stronger in oxygen than in nitrogen flow, increasing the porosity and film inhomogeneity during the pyrolysis step. Upon thermal growth process up to 1000°C, highly textured CeO_2 layers were obtained independently of the composition of the gas flow for heating rates of 10°C/min.

However, we observed that intermediate plateaus or slow heating rates are detrimental for epitaxial reconstruction during the growth process, resulting in polycrystalline films.

These results evidence the importance of adjusting the processing parameters to ensure full epitaxial growth.

ACKNOWLEDGMENTS

We acknowledge the financial support from Spanish MINECO (CONSOLIDER, NANOSELECT, MAT2011-28874-C02-01, IPT-2011-1090-920000, Ramón y Cajal program on behalf of A.C.), Generalitat de Catalunya (Pla de Recerca 2009-SGR-770 and XaRMAE) and EU (Nespa-Marie Curie).

*Corresponding author at: acalleja@icmab.es

REFERENCES

1. A.Trovarelli, Catalysis Review Science and Engineering, **38**, 439 (1996).
2. M. Mogensen, N.M. Sammes, G.A. Tompsett, Solid State Ionics, 129, **63** (2000).
3. B. Harrison, A.F. Diwell, C.Hallet, Platinum Metals Review, **32**, 73 (1988).
4. F.C. Chiu, C. M. Lai. Journal of Physics D: Applied Physics, **43**, 075104 (2010).
5. M. Coll, J. Gazquez, F. Sandiumenge, T. Puig, X. Obradors, J. P. Espinos, R. Hühne, Nanotechnology **2008**, 19, 395601.
6. A. Cavallaro, F. Sandiumenge, J. Gazquez, T. Puig, X. Obradors, J. Arbiol, H.C. Freyhardt, Adv. Funct. Mater. **2006**, 16,(10) 1363-1372.
7. M. Coll, J. Gazquez, R. Hühne, B. Holzapfel, Y. Morilla, J. Garcia-Lopez, A. Pomar, F. Sandiumenge, T. Puig, X. Obradors, J. Mater. Res. **2009**, 24,(4) 1446-1455.
8. Q.X. Jia, T.M. McCleskey, A.K. Burrell, Y. Lin, G. E. Collis, H. Wang, A.D.Q. Li, S.R. Foltyn, Nature Materials, **3**, 529 (2004).
9. M. Jain, E. Bauer, Y. Lin, H. Wang, A.K. Burrell, T.M. McClesky, Q.X. Jia, Integrated Ferroelectrics, **100**, 132–139 (2008).
10. H. M. Luo, M. Jain, S. A. Baily, T. M. McCleskey, A. K. Burrell, E. Bauer, R. F. DePaula, P. C. Dowden, L. Civale, Q. X. Jia, Journal of Physical Chemistry B, **111**, 7497-7500 (2007).
11. A.E. Martell, R.M. Smith, Critical stability constants, vol. 2. Plenum Press, New York (1975).

12. R.M. Smith, A.E. Martell, Critical stability constants, vol. 3. Plenum Press, New York (1977).
13. M. Coll, A. Pomar, T. Puig, X. Obradors, Applied Physics Express, **1,** 121701 (2008).
14. K.C. Lanigan, K. Pidsosny, Vibrational Spectroscopy, **45,** 2–9 (2007).
15. X. Obradors, F. Martínez-Julian, K. Zalamova, V. R. Vlad, A. Pomar, A. Palau, A. Llordés, H. Chen, M. Coll, S. Ricart, N. Mestres, X. Granados, T. Puig, M. Rikel, Physica C, in press, DOI 10.1016/j.physc.2012.04.020.

Mater. Res. Soc. Symp. Proc. Vol. 1449 © 2012 Materials Research Society
DOI: 10.1557/opl.2012.849

Preparation and Characterization of Pb(Zr,Ti)O$_3$ films prepared by a modified sol-gel route

Dan Jiang, Chen Zhao, Shundong Bu, and Jinrong Cheng *
School of Materials Science and Engineering, Shanghai University, Shanghai, 200072, China

ABSTRACT

Pb(Zr$_{0.53}$Ti$_{0.47}$)O$_3$ (PZT) films have been fabricated on stainless steel substrates by a Polyvinylpirrolidone (PVP) modified sol-gel route. The single layer of about 0.26 μm was achieved by using the PVP-modified PZT sol, and Crack-free PZT films with thickness of up to 2.37 μm were fabricated by repeating the deposition process. The variations in crystallite orientation, microstructure, dielectric and ferroelectric properties of PZT films were investigated as a function of film thickness. Our results indicate that PZT films prepared on stainless steel substrates maintain good dielectric and ferroelectric properties.

INTRODUCTION

Up to now, lead zirconate titanate (PZT) have attracted much attention for broader application in piezoelectric actuators, micro-electro-mechanical system (MEMS), and novolatile ferroelectric random access memories (FeRAM) due to its superior dielectric, piezoelectric, and ferroelectric properties. The film thickness of PZT in the range of 0.5-10 μm is likely to be required for full elicitation of such properties [1-3].

Currently, PZT films have been deposited on the metal substrate instead of the platinized silicon wafers. The advantages of PZT films prepared on metal substrates include low cost, high frequency operation, obvious displacement, quick response, and high fracture toughness [4-6].

Among the methods of preparing PZT films, the sol-gel process is widely applied because of the low cost and easy compositional control. However, it is generally difficult to achieve crack-free films with thickness over 1 μm by the conventional sol-gel method. In order to avoid cracking, the single-layer thickness is controlled below 0.1 μm, and repetitive deposition is performed in laboratories, which is impracticable in industries [7-8]. Polyvinylpyrrolidone (PVP) has been reportedly used to fabricate BaTiO$_3$, BaBi$_4$Ti$_4$O$_{15}$, and Pb(Zr,Ti)O$_3$ (PZT) ceramic films with single-layer thickness of 1.2, 0.4, and 0.75 μm, respectively [9-10]. H. Kozuka et al. previously demonstrated that the addition of PVP in alkoxide solutions allows crack-free, over submicrometer-thick ceramic coating films to be obtained via noncycled, single-step deposition [11].

In this work, we have studied a polymer-assisted sol-gel method by the addition of PVP to increase the single-layer thickness of the PZT films. It is found that the single-layer thickness of PZT films increases from 0.09 to 0.26 μm after the PZT precursors modified by PVP. The structure, dielectric and ferroelectric properties of PZT films are studied as a function of film thickness.

EXPERIMENTAL DETAILS

Tetrabutyl titanate [Ti(OC$_4$H$_9$)$_4$], lead acetate trihydrate [Pb(CH$_3$COO)$_2$•3H$_2$O] and tetrabutul zirconate [Zr(OC$_4$H$_9$)$_4$] were used as raw materials. Ethylene glycol monomethylether (CH$_3$OCH$_2$CH$_2$OH, 2-MOE) as used as solvent. The sol was prepared using a PVP modified sol-gel method. PVP of 38,000 in viscosity average molecular weight (k-30) was used. The molar composition of the PZT: PVP=1:1 where the molar ratio for PVP was defined for the monomer (the polymerization unit). Excess Pb of 20 at% was employed to compensate for the loss of lead on firing. The PVP powder was first dissolved into the 2-methoxyethanol, then PZT precursor was prepared by dissolving leadacetate, zirconium isopropoxide and titanium tetrobutoxide into the above 2-methoxyethanol solvent. The solution was stirred at 90 °C for 2 h to get the stabilized yellow sols and the concentration was 0.5 mol/L.

A single-layer PbTiO$_3$ (PT) of about 20 nm was firstly deposited on the SS substrates to reduce the annealing temperature of PZT films. PZT films with coatings of 3, 6, 9 and 12 layers were deposited on the PT buffered SS substrates respectively, with a rapid thermal annealing at 550 °C for 3 min for each layer, he final films were annealed at 650 °C for 30 min in air. The crystal structure of PZT films were obtained by X-ray diffraction (XRD). The morphology of surface was observed by the scanning electron microscopy (SEM). Radiant Technology Ferroelectric tester (Precision Premier II) was used to measure the ferroelectric hysteresis loops and the J-V curve using the step delay and soak time of 100 ms.

RESULTS AND DISCUSSION

The crystalline structure of the PZT films was observed by the x-ray differaction (XRD) analyses. Patterns were recorded at a rate of 4°/min in the 2θ range of 20-80 °. Fig. (1) exhibits XRD patterns of the films with various thicknesses, it shows that PZT films annealed at 650 °C for 30 min formed a pure perovskite structure without any secondary phases. The addition of PVP did not affect the crystal structure of PZT films. No significant variation of the crystalline structure can be observed for the films as the thickness increases, it can be seen that all the films have a similar pattern of the randomized orientation. The (110) diffraction peak of PZT films becomes sharper and stronger as the films thickness increase, suggesting that the higher PZT thickness is beneficial to the growth of the (110) oriented PZT films.

42

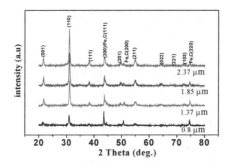

Fig.1. XRD patterns of PZT films for different thickness annealed at 650 °C

Fig. 2 illustrates the surface morphology and cross-section SEM images of the PZT films with different thickness. It was observed that all the films showed a similar surface morphology. From the fig. 2 (a), (b), (c) and (d), the SEM analysis showed a crack-free surface and porous microstructure. There is some small porosity appeared in the surface of the PZT films, however the porosity do not penetrate through the underlayer based on the observation of the cross-section view. Hence it will not cause the following electric problems. It was reported that there were strong hydrogen bonds between the amide groups of PVP and the hydroxyl groups of the metalloxane polymers could hinder the condensation reaction in films, promoting the stress relaxation in the heat-up stage and allowing the films to be crack-free [12]. The fig. 2 (c) and (d) revealed that the films microstructure consisted of round lighted-colored regions (rosettes) and dark regions (between the rosettes). That the rosettes region is rich in Pb and the drak regions deficient in Pb for PZT films, which is in consistent with results reported by Ebru [13]. However, its formation mechanism has not been well understood. Suggested reasons include diffusion of Pb to the bottom electrodes, Pb volatilization during annealing, and flash decomposition of the sol-gel precursors [14].

The estimated thickness of the film with 3, 6, 9 and 12 layers from the SEM micrograph of the cross section were about 0.8, 1.37, 1.85 and 2.37 μm, respectively. Obviously, the addition of PVP increases the viscosity of the solution, resulting in the increase of coating thickness. At the same time PVP suppresses the stress in the thick layer, avoiding the cracking formation in the PZT films.

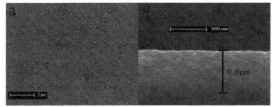

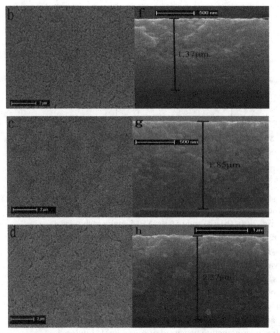

Fig. 2. Surface morphology and cross-section SEM images of the PZT films with different thicknesses: (a, e) 3 L, (b, f) 6 L, (c, g) 9 L and (d, h) 12 L

Fig. 3 presents the relationship of the dielectric constant and the dielectric loss according to the frequency for different films thickness. The dielectric constants of all films gradually decreased with increasing frequency, and were significantly improved with increasing thickness. The dielectric loss shows an opposite variation to that in dielectric constant. The PZT films with the thickness 2.37 μm exhibited excellent dielectric properties, the values of ε_r and tanδ of the films at 10^6 Hz are 910 and 0.07, respectively.

These properties can be explained by the reduction of porosity, as show in the microstructures. It was estimated that about 25%-50% of the dielectric response at room temperature was from extrinsic sources. The extrinsic contribution to the dielectric constant of the PZT films was mainly attributed to 180° domain wall motion, which increased with film thickness. The behavior of the dielectric loss with thickness is still not well understood. A similar mechanism may also be responsible for the dielectric loss behavior in PZT films [15-16].

Fig. 4 shows the curves of the electric current density (J) versus dc field. The Fig.ure shows that the PZT films with thickness of 2.37 μm have the lowest leakage current density of 10^{-8} A/cm^2 under the field of 100 KV/cm.

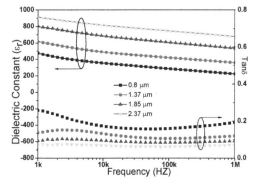

Fig. 3. Dielectric constant and loss as a function of frequency of the PZT films with different thicknesses

Fig. 5 shows the P-E characteristic of PZT films with different thicknesses. It is clear to see that normal ferroelectric behavior was obtained. Although the shape of the PZT films of different thicknesses do not change dramatically, the remnant polarization (Pr) of the PZT films increases with the thicknesss increase. It can be explained by the improved density of the PZT structures. It implies that for increasing number of processing round, a much denser PZT material structure can be obtained. At twelve rounds of coating, the maximum remanent polarization is 68 $\mu C/cm^2$ and the coercive field is 106 KV/cm at the maximum electric field, 250 KV/cm. The addition of PVP, which was able to avoid film cracks, promoted film thickness and decreased the clamping effect from the substrate, resulting good ferroelectric behavior.

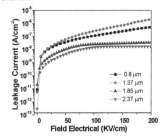

Fig. 4. Leakage current density of PZT films with different thicknesses

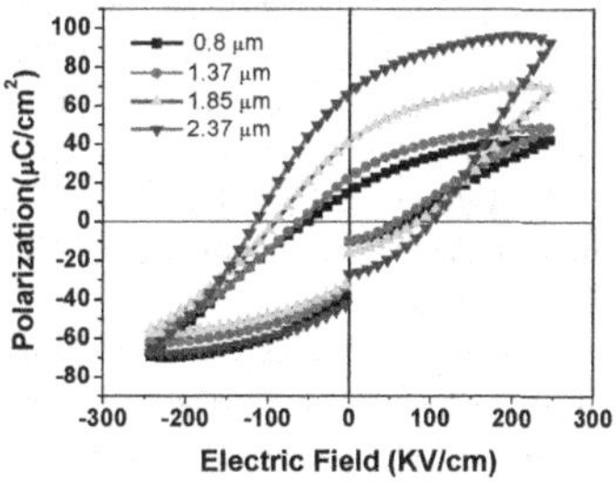

Fig. 5. P-E characteristic of PZT films with different thicknesses

CONCLUSIONS

PZT films prepared from PVP-assisted solutions exhibit good dielectric and ferroelectric properties. When the thickness of the films is 2.37 μm, the maximum remanent polarization is 68 $\mu C/cm^2$ and the coercive field is 106 KV/cm, and its dielectric and ferroelectric properties are 910 and 0.07 respectively. The data shows that dielectric and ferroelectric properties of the PZT films are comparable with those of PZT films fabricated by a conventional sol-gel process, indicating that the PZT films prepared in this way are ready for potential application.

ACKNOWLEDGMENTS

This work was supported by Shanghai education development foundation under grant No. 08SG41 and National Nature Science Foundation of China under Grant No. 50872080.

REFERENCE

[1]B. Willeng, M. Kohli, K. Brooks, P. Muralt and N. Setter, J.Ferroelectrics 201, 147(1997).

[2]Z.H. Zhou, J.M.Xue, W.Z.Li, Wang, H.Zhu, and J.M.Miao, J.Appl.Phys.96, 5706 (2004).

[3]Y.L. Tu and S.J.Milne, "Processing and Characterization of Pb(Zr,Ti)O₃ Films up to 10 μm Thick, Produced from a Diol Sol-Gel Route, " J. Mater. Res.11, 2556 (1996).

[4]So Baba, Hiroki Tsuda, and Jun Akedo J.IEEE Transactions on ultrasonics, ferroelectrics, and frequency control, Vol.55, NO.5, May 2008.

[5]Jae-Hyuk Park, JunAkedo, HarumichiSato, J. Sensors and Actuators A 135, 86 (2007).

[6]Jinrong Cheng, Liang He, Lingjuan Che, Zhongyan Meng, J. Thin Solid Films 515, 2398 (2006)

[7]Akihiro Yamano and Hiromitsu Kozuka, J.Am.Ceram. Soc, 90, 3882 (2007).

[8]Gun-Tae Park, Jong-Jin Choi, Chee-Sung Park,Jae-Wung Lee, and Hyoun-Ee Kim, Applied Physics Letters, volume 85, Number 1220 September 2004.

[9]H. Kozuka, M. Kajimura, T.Hirano, and K.Katayama: Crack-free, thick ceramic coating films via non-repetitive dip-coating using polyvinylpyrrolidone as stress-relaxing agent. J. Sol.-Gel Sci.Technol. 19, 205 (2000).

[10]S.Takenaka and H. Kozuka: Sol-gel preparation of single-layer, 0.75 μm thick lead zirconate titanate films from lead nitrate-tita-nium and zirconium alkoxide solutions containing polyvinylpyrrolidone. Appl.Phys.Lett. 79, 3485 (2001).

[11]H.Kozuka, "Stress Evolution on Gel-to-Ceramic Thin Film Conversion," J.Sol-Gel Sci.Techn, 40, 287 (2006)

[12]Hiromitsukozuka, Masahirokajimura et al. Crack-Free thick Ceramic Coating Films via Non-Repetitive Dip-Coating Using Polyvinylpyrrolidone as Stress-Relaxing Agent, Journal of Sol-Gel Science and Technology 19, 205, 200.

[13] Mensur Alkoy, Sedat Alkoy, and Tadashi Shiosaki, Effects of Ce, Cr and Er Doping and Annealing Conditions on the Microstructural Features and Electrical Properties of PbZrO₃ Thin films prepared by Sol-Gel Process. Japan J. Appl. Phys.44, 6654 (2005)

[14]H.Du, T.S.Zhang, and J.Ma, Effect of polyvinylpyrrolidone on the formation of perovskite phase and rosette-like structure in sol-gel-derive PLZT films, J.Mater.Res, Vol. 22, No.8, Aug 2007.

[15]F.Xu, S. Trolier-McKinstry, W.Ren, Baomin Xu, Z.-L.Xie et al. J.Appl. Phys.89, 1336 (2001).

[16]Perez de la Cruz, E.Joanni, P. M. Vilainho, and A.L.Kholkin, J.Appl. Phys.108, 1141 (2010).

Mater. Res. Soc. Symp. Proc. Vol. 1449 © 2012 Materials Research Society
DOI: 10.1557/opl.2012.1172

Annealing Temperature, Time and Thickness Dependencies in (TCO) SnO₂ Thin Films Grown by Spray Pyrolysis Technique

Alfredo Campos, Amanda Watson, Ildemán Abrego and E. Ching-Prado*
Natural Science Department, Faculty of Science and Technology, Technological University of Panama, Panama.

ABSTRACT

Tin oxide thin films were prepared by spray pyrolysis method using $SnCl_2.2H_2O$ as starting precursor and deposited on glass substrate. Three groups of samples with different preparation conditions (temperature, time and thickness) were synthetized. The samples were characterized using Scanning Electron Microscope (SEM), X-Ray Diffraction (XRD), UV-Visible Spectroscopy and Van der Pauw four-point electrical measurements. The grain size in the samples changes from 80 to 500 nm. Optical and electrical parameters were measured or calculated, such as: band gap, refractive index, sheet resistance, transmittance spectrum and figure of merit. Film thicknesses were obtained from fringes features in the transmittance spectra with a variation from 76 to 761 nm. A mechanism of transformation from tin dichloride to tin oxide is proposed and discussed; additionally the visual yellow color of some samples, related with a low transparency, is associated to the amount of abhurite or tin hydroxide complex coexisting with tin oxide. The figure of merit showed that 500 °C, 42 sprays and 1 hour of annealing time were the best conditions in the preparation of SnO_2 with TCO properties.

INTRODUCTION

Among the wide range of materials with technological applications, deserve special attention materials that are good electrical conductors and transparent. These materials are known as TCOs (transparent and conductive oxides). The interesting properties of these materials allow their use in liquid crystal displays (LCD), light emitting diodes, solar cells, electrical contacts, touch screens and others applications [1]. Different semiconductors have been used as TCOs, among these: SnO_2, CdO, ZnO, Cd_2SnO_4, $CdIn_2O_4$ and In_2O_3 [2].

The most widely used of all these compounds in technological applications is the indium oxide doped with tin (In_2O_3: Sn), known as ITO. The easy deposition, high conductivity and high transparency of the ITO make it stand out above all other semiconductors. Being the Indium expensive and scarce, recent researches have focused on search new alternatives. One of the most promising candidates is the SnO_2 which has good chemical stability and mechanical strength at high temperatures.

The SnO_2 has a rutile tetragonal structure with lattice parameters a = b = 4.7382 Å, c = 3.1871 Å. The big band gap of tin dioxide (3.62 eV at 27 ° C) does not allow transitions between bands when it is exposed to electromagnetic radiation in the visible spectrum, making it transparent. Oxygen vacancies are intrinsic defects that usually appear in this semiconductor. These defects play the role of electron donor impurities, thus increasing the electrical conductivity [3].

SnO$_2$ thin films have been prepared by different techniques, such as: spray pyrolysis, dip-coating, magnetron sputtering, thermal evaporation, chemical vapor deposition (CVD), among others. The spray pyrolysis is simple, low cost, reproducible, easy for introduction of impurities, fast growing, high capacity for mass production of coatings on substrates of different sizes and not requiring vacuum [2]. All these advantages make the spray pyrolysis technique widely used to prepare different thin film coatings.

One of the precursors most commonly used to prepare SnO$_2$ through spray pyrolysis is the SnCl$_4$. A precursor less used is SnCl$_2$ that is cheaper and easy to manipulate, but getting SnO$_2$ from SnCl$_4$ is achieved at lower temperatures than SnCl$_2$ [4].

This investigation aims at the preparation of SnO$_2$ thin films on glass substrates by spray pyrolysis method using SnCl$_2$ precursor and the application of different experimental techniques to study its properties as TCO.

EXPERIMENTAL DETAILS

To prepare the starting solution, 22.55 g SnCl$_2$.2H$_2$O was dissolved in 5 ml of concentrated hydrochloric acid and 50 ml of distilled water. The clear solution obtained was stirred magnetically for 1 hour and then was diluted with distilled water to achieve the concentration 1M. For starting solution atomization a commercial airbrush spray was used with air as carrier gas. A hotplate was used as substrate heater with a PID temperature controller, allowing temperature errors no greater than 5 ° C for all the depositions. The starting solution was sprayed over glass substrates 25.4 x 25.4 x 1mm^3 with a spray time less than one second to avoid cooling effect and 3 minutes of spray interval to allow thermal stabilization. After the last spray, the samples were annealed in air on the hotplate for a defined time (annealing time). The system was optimized at 10.6 ml / min solution flow, 200 kPa, 30 cm atomizer-substrate distance and 45 ° atomizer tilting. In order to get samples with a good quality, the substrates were cleaned using soap, distilled water, acetone and isopropyl alcohol. A scanning electron microscope ZEISS EVO 40 VP was used to study the morphology of the film surfaces. The X-ray diffractograms were obtained using a SIEMENS D5000 diffractometer. For optical characterization of the films, a spectrophotometer UV-Visible ESPECTRONIC GENESIS 5 was utilized and the sheet resistance measurements were performed by four-point Van der Pauw method.

THEORY

Transmittance spectra of transparent thin film usually present fringes (maximum and minimum of intensity). These fringes are due to many reflections that experience the electromagnetic radiation within the thin film interfaces forming an interference pattern. The fringes features allow estimating the thickness of the films.

$$d = \frac{\lambda_1\lambda_2}{2(n_1\lambda_2 - n_2\lambda_1)} \tag{1}$$

Where d is the thickness of the film and n$_1$, n$_2$, λ_1 and λ_2 correspond to refractive indexes (n) and wavelengths (λ) of two maxima or minima consecutives.

A good parameter to study TCO properties is the figure of merit (ϕ) that relates electrical sheet resistance (R_{sh}) and Optical transmittance (T).

$$\phi = \frac{T^{10}}{R_{sh}}$$ (2)

RESULS AND DISCUSSION

Optical properties

Thin films transmittance spectra, recorded from 200 to 1100 nm, are shown in figure 1. Figure 1a shows the samples prepared with 6 sprays, 1 hour annealing and at different deposition temperatures. The lowest values of transmittance (less than 40 %) are observed in the samples annealed at 200 and 300 °C and at higher deposition temperatures the transmittance values are the highest (more than 75 %). Obviously, these results are associated to films transparency. In 200 °C annealing temperature, the film looks opaque and whitish, while at 300 °C the film also looks opaque, but yellowish. For higher annealing temperatures the films are more transparent. Figure 1b corresponds to samples annealed at 500 °C, 1 hour annealing and with different number of sprays. Fringes features are observed in all the spectra and it causes plot intersections after 400 nm. With increasing number of sprays, the number of fringes increase and the transmittance values decrease before 400 nm. In general terms, in the samples annealed at 500°C with 6 sprays and different annealing time, the transmittance spectra behavior are similar (see figure 1c). However, a detailed observation of the spectra before 600 nm reveals the lowest transmittance values for samples at 5 and 15 minutes annealing time. After this wavelength all the spectra have similar values.

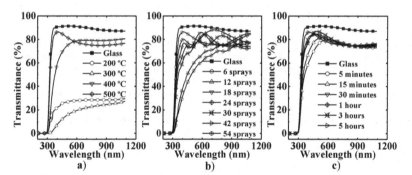

Figure 1. Transmittance spectrums of films. a) 6 sprays, 1 hour annealing and different deposition temperatures. b) 500 °C, 1 hour annealing and different number of sprays. c) 500 °C, 6 sprays and different annealing times.

Thickness of films annealed at 500 °C, 1 hour annealing and different number of sprays was calculated by equation (1). Figure 2 shows a linear regression in thickness behavior with the number of sprays. A very good fitting can be observed with $r^2 = 0.98$.

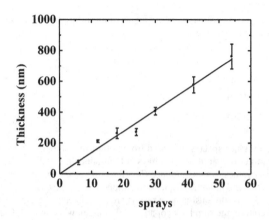

Figure 2. Thickness of films annealed at 500 °C, 1 hour annealing and different number of sprays.

Optoelectronic properties

Figure of merit study was made using Haacke's equation (equation 2). To use a transmittance value in this equation, the transmittance spectra were fitted and the sine part was removed, thus the plot intersections were avoided. This procedure allowed knowing spectrums without fringes features and an average was taken from 580 to 826 nm. The sheet resistance measurements were realized using Van der Pauw method and the electrical contacts were made with silver epoxy. In Table I, optoelectronic properties are presented for samples annealed at 500 °C, 1 hour annealing and different number of sprays. A maximum value of figure of merit was obtained with 42 sprays. In table I, the decrease in average transmittance with increasing number of sprays is shown and it is due to the fact that transmittance depends inversely with thickness. Also it is found a decrease in sheet resistance with increasing number of sprays. The last behavior can be explained as follow: there are different scatter mechanisms that affect the charge carries movement in semiconductors, among these: photon scattering, impurity and defect scattering. In thin films the surface scattering must be considered. If the thickness is greater than the electron free path in the bulk, the surface scattering decrease and the bulk dominate [5].

Table I. Optoelectronic properties of films prepared at 500 °C, 1 hour annealing time and different number of sprays.

Opto-electronic Properties	Number of sprays						
	6	12	18	24	30	42	54
Average T (%)	82.29	81.96	80.00	79.78	78.45	71.49	66.92
R_{sh} (kΩ/sq)	14.82	2.282	1.46	1.58	1.84	0.40	0.24
Φ (10^{-5} Ω^{-1})	0.96	5.99	7.35	6.63	4.80	8.74	7.36

Structural properties

Figure 3a shows the X-ray diffractograms of films with 6 sprays, 1 hour annealing and annealed at different temperatures, where the crystal planes (110), (101), (200) and (211) can be observed. They are characteristic of tetragonal SnO2 structure. It reveals a polycrystalline nature at all samples. At 200 °C most of the peaks belong to $Sn_3O(OH)_2Cl_2$ (abhurite material) and do not appear any evidence of SnO_2 phase. At 300 °C peaks signal corresponding to SnO_2 tetragonal phase can be identified in the x-ray diffractograms, but also appear another group of peaks associated to $Sn_6O_4(OH)_4$ complex. The incomplete formation of SnO_2 at these temperatures seems to be responsible of the opaque coloration in these films (see figure 1a). The disappearance of abhurite to form $Sn_6O_4(OH)_4$ seems to be caused by the displacement of chlorine by oxygen with increasing temperature. At 400 °C the presence of $Sn_6O_4(OH)_4$ decrease and the presence of SnO_2 tetragonal phase is maintained. In this film, the transparency is better than films annealed at 200 and 300 °C. At 500 °C all the peaks correspond to SnO_2 in structure and it looks totally transparent.

Morphological study

All films micrographs confirm polycrystalline structures with nanometric grains (figure 3b). A count of the number of grains and their sizes was made for samples with different deposition temperature, number of sprays and annealing time. It was realized to make grain size histograms. The most appreciable changes in the distributions were observed at different number of sprays with central histograms positions increasing from 130 to 550 nm. Increasing annealing temperature, the grain sizes histograms change from 80 nm to 130 nm. For increasing annealing time, grain sizes from 95 to 180 nm were found. The width of the distributions increases with increasing number of sprays. Also, the distributions were moved at higher grain size values and the grains present irregular form, where the number of sprays increases. The last result is in good agreement with the literature [6].

a)   b)

Figure 3. a) X ray diffractograms of the films annealed at 200, 300, 400 and 500 °C, 6 sprays and 1 hour annealing time. b) SEM image of 30 sprays, 1 hour annealing and annealed at 500 °C film at 50 kx.

CONCLUSIONS

SnO_2 thin films with TCO properties were prepared by spray pyrolysis technique and using $SnCl_2$ as starting material. High optical transmittance values (>75 %) were achieved with increasing annealing temperature, while transmittance values decrease with increasing number of spray (thickness). To different annealing time the behavior of transmittance spectrums are similar. The thickness of the films changes linearly with the number of sprays; from 76 nm at 6 sprays to 761 nm at 54 sprays with a rate of 13.8 nm/spray. This indicates a good control in the growth of the films. The formation of SnO_2, phase starting from $SnCl_2$, is mediated by two intermediate phases. The first one is called abhurite ($Sn_3O(OH)_2Cl_2$); and the second one is a tin hydroxide complex ($Sn_6O_4(OH)_4$). At 500 °C only appear SnO_2 tetragonal phase and all the complex $Sn_6O_4(OH)_4$ react to form SnO_2. For this reason, 500 °C is the best temperature to prepare SnO_2 from this starting solution. The best value of figure of merit (8.7410^{-5} Ω^{-1}) was achieved at 500 °C annealing temperature, 1 hour annealing and 42 sprays. This is the best condition to prepare TCO tin oxide thin films using $SnCl_2$ precursor. Good grains structures were obtained in the three groups of samples, all of them are nanometric order (from 80 to 550 nm).

ACKNOWLEDGMENTS

The authors are grateful to SENACYT (projects FID 05-061 and APY-GC-10-046A) for financial support. All the authors thank to SmithSonian Tropical Research Institute and in special to Jorge Ceballos who helped us measuring, analyzing and discussing the morphological characterization (SEM). Our sincere thank to the Institute of Materials Jean Rouxel for recording the X-ray diffractograms.

*Corresponding author E-mail address: eleicer.ching@utp.ac.pa.

REFERENCES

1. F. Xsiaocheng, Y. Yian, H Linfeng, L. Hui and L. See, *Advanced Functional Materials*, 22(8) 1613-1622, 2012.
2. E. Elangovan and K. Ramamurthi, *Journal of Optoelectronics and Advanced Materials*, 5(1), 45-54, 2003.
3. S. Ray, P. Gupta and G. Singh, *Journal of Ovonic Research*, 6(1), 23-34, 2010.
4. H. García and A. Martínez, *Recent Advances in Circuits, Systems, Signal and Telecommunications*, 144-146, 2010.
5. F. Amanullah, M. Mobarak, A. Dhafiri and K. Shibani, *Materials Chemistry and Physics*, 59, 247-253, 1999.
6. R. Fernández, PhD. Thesis, University of Habana, 2005.
7. G. Cediel, F. Rojas, H. Infante and G. Gordillo, *Colombian Magazine of Physics*, 34(1), 48-54, 2002.
8. A. Tricoli, M. Righettoni and A. Teleki, *Semiconductor Sensor*, 7632-7659, 2010.
9. L. Huiyong, N. Avrutin, Ü. Izymskaka, Ü. Özgür and H. Morkoç, *Superlattices and Microstructure*, 48, 458-484, 2010.
10. K. Ravichandran, G. Murugananthem, B. Sakthivel and P. Philominathan, *Journal of Ovonic Research*, 5(3), 63-69, 2009.

Mater. Res. Soc. Symp. Proc. Vol. 1449 © 2012 Materials Research Society
DOI: 10.1557/opl.2012.793

Fabrication and electrical properties of $0.7BiFeO_3$-$0.3PbTiO_3$ films on stainless steel by the sol-gel method

Chen Zhao, Dan Jiang, Shundong Bu and Jinrong Cheng
School of Materials Science and Engineering, Shanghai University, Shanghai 200072, China

ABSTRACT

Ferroelectric $0.7BiFeO_3$-$0.3PbTiO_3$ (BFO-PT) films were deposited on stainless steel substrates by the sol-gel method. A thin layer of $PbTiO_3$ (PT) was introduced between the substrates and BFO-PT films in order to decrease the annealing temperature of BFO-PT films. X-ray diffraction analysis reveals that BFO-PT films could be well crystallized into the perovskite structure at about 575 °C. Scanning electron microscope (SEM) images show that BFO-PT thin films have grain size of about 50~60 nm. Our results indicated BFO-PT films deposited on stainless steel substrates maintained the excellent ferroelectric properties with remnant polarization of about 40~50 $\mu C/cm^2$.

INTRODUCTION

$xBiFeO_3$-$(1-x)PbTiO_3$ (BFO-PT) solid solutions have attract much attention due that a large c/a ratio of more than 1.18 and high Curie temperature of 623 °C could be achieved in the vicinity of morphotropic phase boundary (MPB) of x≈0.7 [1]. The excellent piezoelectric and ferroelectric properties of bulk BFO-PT are expected to be duplicated in BFO-PT thin films for potential applications of micro electro mechanical system (MEMS). It has been reported the La modified BFO-PT thin films deposited on platinized Si substrates exhibit good dielectric and ferroelectric properties [2]. The obvious photo-electric responses have also been observed in BFO-PT films on indium tin oxide (ITO) substrates [3]. The most present research indicated that BFO-PT films are excellent multifunctional materials integrating electrical, optical and magnetic properties simultaneously.

Ferroelectric thin films deposited on base metal substrates have some advantages over the films on conventional Si substrates, which are beneficial to integrating the functional films with engineering system [4-5]. Base metal could serve as both substrate and electrode, allowing a easy process for device fabrication by avoiding the complex micro-machined processes of Si based materials. $Pb(Zr,Ti)O_3$ (PZT) thin films with excellent ferroelectric and piezoelectric properties have been successfully deposited on Ti, Ni, and stainless steel (SS) substrates [6-8]. Addition of a buffer layer, such as $PbTiO_3$, conductive $LaNiO_3$ et al. plays an important role to balance the thermal and lattice mismatch between the films and metal [9]. It is nature of us to prepare BFO-PT films on metal exploring the novel multifunctional properties of such integration.

In this work, BFO-PT films were prepared on a PT-coated stainless steel substrates by the sol-gel method. The annealing temperature and electrical properties of BFO-PT films for different thicknesses were characterized and investigated.

EXPERIMENT DETAILS

0.7BiFeO$_3$-0.3PbTiO$_3$ (BFO-PT) thin films were fabricated on stainless steel substrates by the sol-gel method. The raw materials were bismuth nitrate pentahydrate [Bi(NO$_3$)$_3$·5H$_2$O], iron nitrate nonahydrate [Fe(NO$_3$)$_3$·9H$_2$O], lead acetate [Pb(C$_2$H$_4$O$_2$)$_2$], and tetrabutyl titanate [Ti(OC$_4$H$_9$)$_4$]. Excess lead of 10 at% was batched to compensate the evaporation of lead during annealing process. 2-methoxyethanol (2-MOE) was applied as the solvent. In order to get highly flawless and dense films, formamide and acetic acid were adopted as the drying control chemical additive (DCCA) and the catalyst, respectively. The BFO-PT solution of 0.4 M was deposited on stainless steel (SS) substrates by the spin coating process and the final films were annealed in air at different temperatures for 30 min. Prior to the deposition of BFO-PT, one PT layer of about 20 nm was introduced between BFO-PT films and SS to decrease the annealing temperature of BFO-PT films.

The phase structure of BFO-PT thin films were characterized by a X-ray diffraction (XRD) meter(DLMAX-2550). The morphologies of the films were observed by scanning electronic microscope (SEM). The Pt electrodes with diameter of 0.4 mm were sputtered on the top of BFO-PT thin films to fabricate the metal-insulator-metal (MIM) capacitor for electrical measurements. The dielectric, ferroelectric and leakage current properties were measured by Aglient 4294A Impedance Analyzer and Radiant Technology Ferroelectric tester (Precision Premier II), respectively.

RESULTS AND DISCUSSION

Figure 1(a) exhibits XRD patterns of BFO-PT thin films annealed at different temperatures for 30 min. It shows that all films were of perovskite structured without detected impure phase, and better crystallinity of BFO-PT films was obtained with increasing the annealing temperature. However, an extra phase of Fe (110) was detected at 2θ of about 44.6 °, indicating the relatively strong diffusion of element Fe from the stainless steel substrate to during annealing at the temperature of 600 ℃. Figure 1(b) presents XRD patterns of BFO-PT thin films with different thicknesses annealed at 575 ℃. It can be seen that BFO-PT thin films exhibits a good crystalline quality with pure perovskite phases. With increasing the thickness of BFO-PT, the (110) and (111) diffraction peaks become stronger while the (100) ones has no obvious change. It also displays that diffusion of Fe was well controlled at 575 ℃ compared with that occurred at 600 ℃.

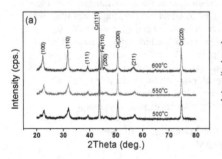
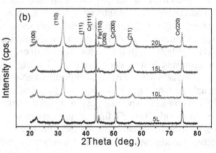

Figure 1. XRD patterns of BFO-PT thin films **(a)** at different annealing temperatures and **(b)** with different thicknesses annealed at 575 ℃

Figure 2 shows SEM images of BFO-PT films with various thicknesses. The BFO-PT films of Fig.2**a, 2c,2e and 2g** were deposited on SS substrates to observe the surface morphologies, while Fig.2**b, 2d, 2f and 2h** represent samples fabricated on Si substrates to give the corresponding cross-sectional images. It can be seen that BFO-PT films exhibit a dense structure without detectable flaws, and the average grain size of BFO-PT films is about of 50~60 nm, being independent on film thickness. A compact cross-sectional images reveal that BFO-PT films are free of cracks with thickness of about 170, 340, 500 and 660 nm for the coating layers of 5, 10, 15 and 20 L respectively.

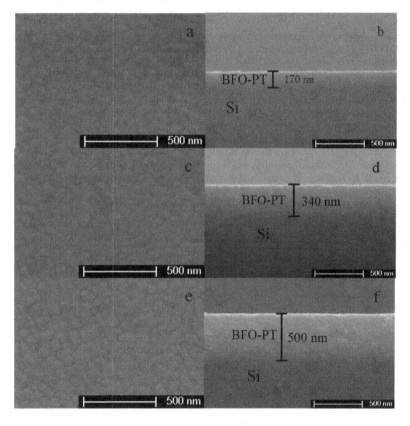

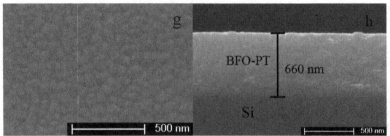

Fiugre 2. SEM images of the surface and cross-sectional BFO-PT films with different thicknesses: (**a,b**) 5 L, (**c,d**)10 L, (**e,f**) 15 L and (**g,h**) 20 L

Figure 3 shows the room temperature relative dielectric constant (ε_r) and dielectric loss (tan δ) of 0.7BFO-0.3PT thin films with different thickness as a function of frequency. It indicates that the dielectric constants of BFO-PT films increase, while the dielectric losses decrease as the film thickness increases. The BFO-PT films with 20 layers have relatively higher ε_r and lower tan δ of about 550 and 0.145 respectively, at the frequency of 1 kHz. The improved dielectric properties of relatively thick BFO-PT films may be due to the mitigated effect of the interfacial layer with increasing the film thickness [10]. Both the values of ε_r and tan δ have a clearly dependence on frequency. The dielectric constants of BFO-PT films decrease gradually with increasing the frequency revealing a dielectric dispersion, which becomes more prominent for relatively thick films. The dielectric loss decreases slightly with frequency at the frequency of less than 20 kHz, and then increases with frequency due to the dielectric relaxation. The higher dielectric loss at lower frequency may result from the leakage conductivity of BFO-PT films, which was improved with increasing the film thickness.

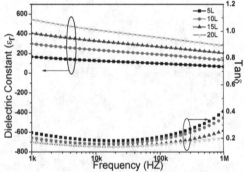

Figure 3. Frquency dependent dielectric constant (ε_r) and dielectric loss (tan δ) of 0.7BFO-0.3PT thin films with different thicknesses

Figure 4 shows the polarization-electrical field (P-E) hysteresis curves of the 0.7BFO-0.3PT films with various thicknesses under electric fields of about 600 kV/cm. The values of Pr are about 34, 43, 49 and 57 $\mu C/cm^2$ for BFO-PT films of 5, 10, 15, 20 layers respectively. The corresponding coercive field Ec are of 377, 264, 247 and 228 kV/cm. It reveals that Pr increases and Ec decreases as the film thickness increases. The well shaped and saturated P-E loop for BFO-PT films of 20 Layers reflects the excellent ferroelectric properties, which are in agreement with their lower dielectric loss. The Pr of BFO-PT films on stainless steel obtained in this work could be higher than that of about 36 $\mu C/cm^2$ for the same BFO-PT filmson Pt/Ti/SiO$_2$/Si substrate [11]. With increasing the film thickness, the movement of Ti^{4+} and O^{2-} ions become more active, and the reverse of spontaneous polarization become easier [12, 13].

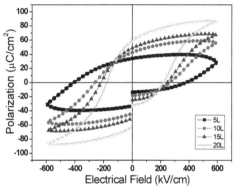

Figure 4. P-E hysteresis curves of BFO-PT films with various thicknesses at room temperature

Figure 5 presents the leakage current density as a function of applied field for BFO-PT thin films with different thicknesses. The BFO-PT film of 20 layers possess the lowest leakage current density of about $10^{-9} \sim 10^{-8}$ A/cm^2 for the applied field from $100 \sim 300$ kV/cm. As the thickness decreases, BFO-PT films tend to display a higher leakage current density, which is in accordance with the variation of dielectric loss with film thickness.

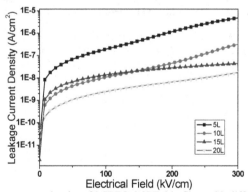

Figure 5. Leakage current density of 0.7BFO-0.3PT thin films with different thicknesses

CONCLUSIONS

The perovskite $0.7BiFeO_3$-$0.3PbTiO_3$ films of different thicknesses were deposited on stainless steel substrates by the sol-gel method under the annealing temperature of 575 °C for 30 min. The dielectric and ferroelectric properties of BFO-PT films increase with increasing the film thickness. BFO-PT films of 20 layers have the largest remnant polarization and the lowest leakage current density of 57 $\mu C/cm^2$ and $10^{-9} \sim 10^{-8}$ A/cm^2 respectively. Our results indicate that BFO-PT films maintain good dielectric and ferroelectric properties.

ACKNOWLEDGMENTS

This work was supported by Shanghai education development foundation under grant No. 08SG41 and National Nature Science Foundation of China under Grant No. 50872080.

REFERENCES

[1] V.V.S.S. Sai Sunder, A. Halliyal, A.M. Umarji, J. Mater. Res., 10, 1301 (1995).

[2] J.R. Cheng, L.E. Cross, Mat. Res. Soc. Proc., 2004. 784, C8.16.1 (2004).

[3] X.W. Zhou, S.W. Yu, B.R. Yuan, W.F. Yang, and J.R. Cheng, IEEE Int. Symp. Appl. Ferroelectric., 2009. Proceedings of the the 2009 18th IEEE International Symposium on Application of Ferroelectrics, 194-197 (2009).

[4] R. Bouregba, G. Poullain, B. Vilquin, H. Murray, Materials Research Bulletin., 35, 1381 (2000).

[5] M. Okada, K. Tominaga, T. Araki, S. Katayama, and Y. Sakashita, Jpn. J. Appl. Phys., 29, 718 (1990).

[6] D.J. You, W.W. Jung, S.K. Choi, and Yasuo Cho, Appl. Phys. Lett., 84, 3346 (2004).

[7] Z.L. Hea, Y.G. Wang and K. Bia, Solid State Communications., 150, 1837 (2010).

[8] R. Seveno, P. Limousin, D. Averty, J.-L Chartier, R le Bihan, H.W. Gundel, J. Euro. Ceram. Soc., 20(12), 2025 (2000).

[9] J.R. Cheng, W.Y. Zhu, N. Li and L.E. Cross, J. Appl. Phys. Lett., 81, 25 (2002).

[10] G.S. Wang, D. Rémiens, E. Dogheche, and R. Herdier, J.Am.Ceram.soc., 89(11), 3417 (2006).

[11] B.R. Yuan, S.W. Yu, W.F. Yang, X.W. Zhou, and J.R. Cheng, IEEE Int. Symp. Appl. Ferroelectric., 2009. Proceedings of the the 2009 18th IEEE International Symposium on Application of Ferroelectrics, 314 (2009).

[12] Q.L. Zhao, M.S. Cao, J. Yuan, R. Lu, D.W. Wang, D.Q. Zhang, Materials Letters., 64, 632-635 (2010).

[13] V. Nagarajan, IG Jenkins, SP Alpay, H. Li, S. Aggarwal, L. Salamanca-Riba, A.L. Roytburd and R. Ramesh, J. Appl. Phys., 86, 595 (1999).

Nanostructures, Nanorods, and Solar or Gas Sensing Applications

Mater. Res. Soc. Symp. Proc. Vol. 1449 © 2012 Materials Research Society
DOI: 10.1557/opl.2012.958

Mg-induced Enhancement of ZnO Optical Properties via Electrochemical Processing

Hongtao Shi, Kalie R. Barrera, Timothy L. Hessong, and Cristhyan F. Alfaro

Department of Physics and Astronomy, Sonoma State University, Rohnert Park, CA 94928, U.S.A.

ABSTRACT

Electrochemical deposition was used to fabricate polycrystalline ZnO thin films in solutions containing zinc nitrate and hexamethylenetetramine. All samples showed intense UV photoluminescence (PL) near the band edge in addition to weak broad bands due to defects. When the source solution was slightly doped with Mg^{2+} ions, the defect induced emission was significantly suppressed while the UV peak position and intensity remained the same. Auger electron spectroscopy revealed no Mg contents in the films within the detection limit. A possible growth mechanism was proposed, based on the chemical reactivity of Mg and Zn, to interpret the observed PL data, which is supported by samples grown in Ca-doped solutions.

INTRODUCTION

Zinc oxide (ZnO) has recently drawn much attention due to its outstanding physical properties and potential applications in fields such as ultraviolet (UV) optoelectronics and photonics [1]. It is generally accepted that the UV photoluminescence (PL) from ZnO is due to the excitonic recombination, whereas a broad band PL involves different types of defects, presenting an obstacle to the realization of practical devices. While most sample fabrication techniques require a growth temperature of 350 °C or higher, a low temperature solution route has indeed demonstrated wafer-scale production of ZnO nanowires [2,3], and possible room temperature ferromagnetism [4]. Moreover, $Mg_xZn_{1-x}O$ alloys have been demonstrated to have a tunable band gap from 3.3 eV to as high as 7.8 eV, which depends linearly on the Mg content in the alloy and could be used in deep UV applications [5,6,7]. While the band gap enlargement is well established in $Mg_xZn_{1-x}O$, few reports have concentrated on the effects of Mg as a dopant on the formation of deep level defects, in particular in a low-temperature solution process, as these defects can significantly affect the optical and electrical performances of ZnO-based devices.

In this paper, we report on the growth of ZnO thin films on aluminum (Al) foils using electrochemical deposition (ECD). Samples, grown in solutions containing Mg^{2+} or Ca^{2+} ions, show much enhanced optical properties, which is attributed to the chemical reactivity of these elements.

EXPERIMENTAL DETAILS

ZnO thin films were grown in a Virgin Teflon cell on 0.2 mm thick, 99.9995% pure Al foils, which were electropolished in a perchloric acid and ethanol mixture (1:4, v/v). In a typical cycle, a solution of 40 ml, containing 20 millimolar (mM) zinc nitrate hexahydrate $[Zn(NO_3)_2 \cdot 6H_2O]$

and 20 mM hexamethylenetetramine ($C_6H_{12}N_4$ or HMT), was used as electrolyte with the latter providing a basic environment, resulting in crystals with well-defined smooth surfaces [8].

A pure ZnO sample (referred to as ZnO:Pure hereinafter) was grown at 80 °C for 3 hours under a constant current density of 0.5 mA/cm². The Al foil and a thin platinum wire were used as the cathode and anode, respectively. To study how impurities in the source solutions affect the optical properties of ZnO, three other samples were prepared after adding 0.8 mM magnesium nitrate hexahydrate [$Mg(NO_3)_2\cdot6H_2O$], 1.6 mM $Mg(NO_3)_2\cdot6H_2O$, and 0.8 mM calcium nitrate tetrahydrate [$Ca(NO_3)_2\cdot4H_2O$] into the zinc nitrate and HMT solutions, labeled as ZnO:Mg4, ZnO:Mg8, and ZnO:Ca4, respectively. All other deposition parameters were exactly the same as for making the ZnO:Pure sample. A Rigaku X-ray diffractometer (XRD) was used to investigate the crystal structure. Temperature dependent PL was measured in a closed-cycle optical cryostat in the temperature range of 20 – 200 K, which was excited by a YAG laser at 355 nm and probed by an Ocean Optics spectrometer with a resolution of 1.5 nm. A Physical Electronics Auger electron spectrometer (AES) was employed to probe the chemical composition of each sample.

DISCUSSION

Figure 1 shows the PL spectra of ZnO:Pure, ZnO:Mg4, and ZnO:Mg8 taken at 20 K. All samples show intense UV peaks near 3.3 eV, which is very close to the band edge of ZnO, as well as broad band emissions, similar to what has been observed in other ZnO samples, prepared, for instance, on conductive glass substrates by ECD [9]. In addition, it can be seen that ZnO:Mg4 and ZnO:Mg8 have much weaker emissions in the visible range than the ZnO:Pure sample, implying that the defect induced PL was significantly reduced when Mg^{2+} ions were added to the solutions. No further improvement was observed with more Mg^{2+} doping in the source solutions.

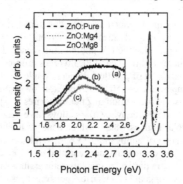

Figure 1. PL spectra from (a) ZnO:Pure, (b) ZnO:Mg4, and (c) ZnO:Mg8 at 20 K. The inset takes a closer look at the data from 1.6 to 2.6 eV.

Figure 2 shows the XRD patterns of the three samples, indicating that all of them are polycrystalline. The following reactions are responsible for the presence of ZnO and $Zn(OH)_2$ peaks due to the fact that the Al substrates were negatively biased during the deposition and that Zn^{2+} is known to have a small standard reduction potential ($E°(Mg) = -0.763$ V) [10,11]:

$$Zn^{2+} + 2H_2O + 2e \rightarrow Zn(OH)_2 + H_2 \qquad (1)$$

$$Zn(OH)_2 \rightarrow ZnO + H_2O \qquad (2)$$

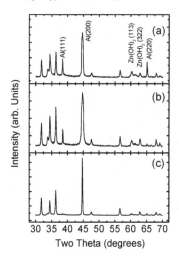

Figure 2. X-ray diffraction patterns of (a) ZnO:Pure, (b) ZnO:Mg4 (b), and (c) ZnO:Mg8. All unlabeled peaks can be undoubtedly assigned to ZnO.

This result is similar to electrolyzing a solution of sodium bicarbonate with a zinc anode or growing ZnO using vapor phase epitaxy [12] except that our growth temperature was not high enough to completely decompose $Zn(OH)_2$ to ZnO and H_2O. As can be seen, the XRD data in figure 2 had little dependence on the Mg-doping in the solutions. No magnesium related phases were discovered in ZnO:Mg4 and ZnO:Mg8, indicating that Mg^{2+} ions did not noticeably affect the degree of crystallinity of the obtained ZnO samples.

To reveal the PL dependence on the measuring temperature and Mg-doping in the source solutions, all PL spectra were deconvoluted into three Lorentz peaks to resolve each individual constituent, representing the laser excitation at 355 nm, the near-band-edge UV emission, and the defect induced broad band, respectively. The sum of the three components agreed very well with our data. While Gaussian fits are often seen in ZnO [7,8], Lorentz distribution is not rare either, depending on the fabrication method and growth parameters [13].

Figure 3 shows the UV peak positions of PL spectra, after such fitting procedure, as a function of measuring temperature in the range of 20 – 200 K. The temperature dependence of each sample's UV peak position was then fitted to the modified Varshni model [14]:

$$E_x(T) = E_x(0) - \alpha T^4 / (T+\beta)^3, \qquad (3)$$

where $E_x(0)$, α, and β are fitting parameters, which are summarized in Table I.

It can be seen from figure 3 and Table I that all three samples have very similar UV peak dependence on temperature, indicating that these emissions are of the same origin, presumably due to the excitonic recombination [7,15]. Within the experimental error, they all have the same E_x when T = 0. The peak position did not show any blue shift as Mg^{2+} ions were added to the solutions, though our original intention was to grow $Mg_xZn_{1-x}O$ alloys. Our β values in these samples are close to 80 K, lower than what other groups have reported [7,15], which can be attributed to a large lattice thermal expansion when samples were grown on aluminum foils [16].

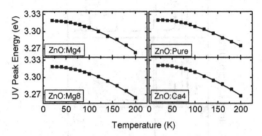

Figure 3. UV peak positions as a function of temperature. Solid lines are fits to equation 3, using near-band-edge data from each sample.

Table I. Summary of the fitting parameters $E_x(0)$, α, and β for ZnO:Pure, ZnO:Mg4, ZnO:Mg8, and ZnO:Ca4, using equation 3.

Sample	E_x (0) (eV)	α (10^{-4} eV/K)	β (K)
ZnO:Pure	3.321	6.0	73.9 ± 7.0
ZnO:Mg4	3.319	8.7	94.2 ± 10.1
ZnO:Mg8	3.318	8.5	95.8 ± 8.2
ZnO:Ca4	3.321	7.5	83.9 ± 7.1

The band gap of $Mg_xZn_{1-x}O$ increases linearly up to 4.15 eV for $0 < x < 0.36$ [5]. In our work, the molar concentration ratios of Mg to Zn in the solutions are 0.04 and 0.08 for ZnO:Mg4 and ZnO:Mg8, respectively, close to the equilibrium solubility limit (4%) of MgO in bulk ZnO [15], but significantly below the maximum values reached by other nonequilibrium growth techniques [18]. If Mg had been introduced into these samples, a blue shift of the UV peak energy should have been observed. As shown in figure 3 and Table I, this clearly is not the case.

To better understand the chemical composition in each sample, first derivative Auger electron spectroscopy was applied, which has a high sensitivity as lighter atoms are known to have a higher Auger yield [19]. Figure 4 shows the data from ZnO:Pure, ZnO:Mg4, and ZnO:Mg8, as well as the well-established elemental positions of oxygen, zinc, and aluminum. Ion beam sputtering in Ar atmosphere was also performed to each sample at a gun voltage of 5 kV, which resulted in almost identical spectra. These measurements therefore verified that no Mg was present in any of these samples within the detection limit.

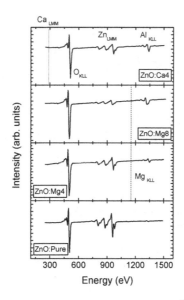

Figure 4. AES spectra show no Mg or Ca in samples doped with Mg^{2+} or Ca^{2+} in solutions.

It is well known that magnesium is higher on the activity series of elements than zinc [20] with a more negative reducing potential (E^{o} (Mg) = -2.375 V) [11]. Due to the addition of magnesium nitrate, following overall reactions likely took place near the deposition zone:

$$Mg^{2+} + H_2O + 2e \rightarrow MgO + H_2 \qquad (4)$$

$$Zn^{2+} + MgO \rightarrow ZnO + Mg^{2+} \qquad (5)$$

The formation of ZnO in equation 5 therefore favored the release of Mg^{2+} back to the solution to produce MgO (equation 4), which facilitated the deposition of more ZnO. Such a process is therefore completely different from equations 1 and 2. To support this proposed mechanism, we prepared a sample in zinc nitrate and HMT solution which contained 0.8 mM calcium nitrate as the impurity instead. Calcium is one period below Mg in the periodic table [20] with a slightly higher negative reduction potential (E^{o} (Ca) = -2.76 V) [11], therefore no calcium should be expected in such a sample. This can be clearly seen in figure 4 that no Ca was discovered. On the other hand, figure 3 and Table I demonstrated that ZnO:Ca4 had very similar PL characteristics to ZnO:Pure, ZnO:Mg4 and ZnO:Mg8. No blue shift of the UV peak position in ZnO:Ca4 was observed either.

CONCLUSIONS

In summary, ZnO thin films were fabricated using electrodeposition. Our preliminary results show that the optical properties of ZnO were significantly enhanced because of Mg-doping in the solutions, which has been attributed to the chemical property of Mg^{2+} in water. Such an enhancement has been observed from 20 to 200 K. To truly reveal the correlation between the PL improvement and Mg-doping in our samples, we need perform further measurements such as transmission electron microscopy or secondary ion mass spectroscopy. This solution process could be applied to other compound semiconductors to enhance their optical properties as well.

ACKNOWLEDGEMENTS

This work was supported by the Sonoma State University Office of Research and Sponsored Programs. We thank Steve Anderson, Nels Worden, and John Collins for their technical help. H.S. is grateful to Profs. Joseph Tenn and Meng-Chih Su for fruitful discussion.

REFERENCES

1. For a review of ZnO, see Ü. Özgür, Ya. I. Alivov, C. Liu, A. Teke, M. A. Reshchikov, S. Doğan, V. Avrutin, S.-J. Cho, and H. Morkoç, J. Appl. Phys. **98**, 041301 (2005).
2. M. H. Huang, S. Mao, H. Feick, H. Yan, H. Kind, E. Weber, R. Russo, and P. Yang, Science **292**, 1897 (2001).
3. Y. Wei, W. Wu, R. Guo, D. Yuan, S. Das, and Z. Wang, Nano Lett. **10**, 3414 (2010).
4. J. B. Cui and U. J. Gibson, Appl. Phys. Lett. **87**, 133108 (2005).
5. A. Ohtomo, M. Kawasaki, T. Koida, K. Masubuchi, H. Koinuma, Y. Sakurai, Y. Yoshida, T. Yasuda, and Y. Segawa, Appl. Phys. Lett. **72**, 2466 (1998).
6. M. Lorenz, E. M. Kaidashev, A. Rahm, Th. Nobis, J. Lenzner, G. Wagner, D. Spemann, H. Hochmuth, and M. Grundmann, Appl. Phys. Lett. **86**, 143113 (2005).
7. M. Trunk, V. Venkatachalapathy, A. Galeckas, and A. Yu. Kuznetsov, Appl. Phys. Lett. **97**, 211901 (2010).
8. J. B. Cui, J. Phys. Chem. C **112**, 10385 (2008).
9. B. Cao, W. Cai, H. Zeng, and G. Duan, J. Appl. Phys. **99**, 073516 (2006).
10. M. Izaki and T. Omi, J. Electrochem. Soc. **143**, L53 (1996).
11. A. Bard, L. Faulkner, "Electrochemical Methods – Fundamentals and Applications," (John Wiley and Sons, Inc., 1980) pp. 699.
12. M. V. Chukichev, B. M. Ataev, V. V. Mamedov, Ya. Alivov, and I. I. Khodos, Semiconductors **36**, 1052 (2002).
13. E. Kärber, T. Raadik, T. Dedova, J. Krustok, A. Mere, V. Mikli, and M. Krunks, Nanoscale Res. Lett. **6**, 359 (2011).
14. Y. P. Varshni, Physica **34**, 149 (1967).
15. Y. S. Jung, W. K. Choia, O. V. Kononenko, and G. N. Panin, J. Appl. Phys. **99**, 013502 (2006).
16. Y. K. Vekilov, A. P. Rusakov, and Fiz. Tverd, Sov. Phys. Solid State **13**, 956 (1972).
17. Y. Li, R. Deng, B. Yao, G. Xing, D. Wang, and T. Wu, Appl. Phys. Lett. **97**, 102506 (2010).
18. A. Ohtomo and A. Tuskazati, Semicond. Sci. Technol. **20**, S1 (2005).
19. K. Siegbahn, "ESCA Applied to Free Molecule," (North Holland, Amsterdam, 1969).
20. F. A. Cotton, G. Wilkinson, Adv. Inorg. Chem. (Wiley-Interscience, New York, 1980) pp. 281-282.

Mater. Res. Soc. Symp. Proc. Vol. 1449 © 2012 Materials Research Society
DOI: 10.1557/opl.2012.921

Gold-Doped Oxide Nanocomposites Prepared by Two Solution Methods and Their Gas-Sensing Response

Chien-Tsung Wang*, Huan-Yu Chen and Yu-Chung Chen
Department of Chemical and Materials Engineering, National Yunlin University of Science and Technology, 123 University Road, Section 3, Douliou, Yunlin, 640, Taiwan

ABSTRACT

Gold species on an oxide support possess variable electronic structures via charge transition so as to increase their chemical redox activity. They are also viably promising for use to enhance gas-sensing response when being exploited in a solid state gas sensor. The synthesis method of the gold-loaded materials plays a crucial role in the functionality. In this paper, we report two types of gold/tin oxide based nanopowders prepared by co-precipitation method and by deposition-precipitation method, respectively. They were evaluated as sensing elements in a semiconductor carbon monoxide (CO) gas sensor. Effects of the material type and CO concentration on sensor response were investigated. Their structural characterizations were done by X-ray photoelectron spectroscopy, X-ray diffraction and transmission electron microscopy. Results demonstrate the surface gold species effective to facilitate CO oxidation in gas atmosphere and promote low-temperature sensor performance.

INTRODUCTION

Semi-conducting metal oxides have been widely exploited as sensing elements in solid-state sensors, because they possess active sites to chemisorb molecules and catalyze reactions. For example, tin dioxide (SnO_2) is most frequently used in gas sensors to detect hazardous species such as carbon monoxide (CO), ethanol, ..etc. This n-type semiconductor adsorbs oxygen in air and CO molecules on the surface, and the oxidation of both species produces carbon dioxide and electrons. A decrease in the electrical resistance represents the sensor response. However, the SnO_2 sensors often operate at higher temperature (e.g., up to 400 °C) in order to achieve a higher sensing response. Unfortunately, this thermal effect frequently causes material instability and high power consumption [1].

Great progress has been made by many research teams with invention of noble metal/oxide nanocomposite materials in powder form. Among the examples, gold nanoparticles or clusters deposited on an oxide support can promote sensor performance, by reducing operating temperature or increasing response magnitude [2,3]. This kinetic effect arises from a high catalytic activity over the gold particles (in form of sphere or semi-sphere) located on the oxide support for CO oxidation [4]. The catalytic properties of the small gold clusters or nanoparticles supported are related to their capability to chemisorb small molecules at defect sites on the gold surface or at the hetero-junction between the gold and the oxide support [5]. The synthesis approach of the gold/oxide composite materials plays a crucial role to influence their structural and chemical properties. For example, a previous study using deposition-precipitation (DP) and traditional co-precipitation (CP) techniques to prepare Au/Fe_2O_3 powders reports a higher gold content in the composite materials by the DP method, since this synthetic route achieves more complete precipitation of the gold species from solution [6].

The present study aimed to prepare gold/tin oxide nanocomposite powders by two solution methods (conventional co-precipitation (CP) vs. deposition-precipitation (DP)) and to compare their structural and CO gas-sensing properties. Effects of the material type and CO concentration on sensor response were investigated. Results are quite encouraging as the work will show below, and demonstrate that the gold dopant is promising for use to enhance sensor response to CO gas at low temperature.

EXPERIMENTAL

The Au/SnO_2 composite powders for use in this work were synthesized by two solution methods. In the co-precipitation approach, tin chloride pentahydrate ($SnCl_4 \cdot 5H_2O$, Aldrich 98%) and gold acetate ($Au(CH_3COO)_3$, Alfa Aesar 99.9%) were used as the precursors and dissolved in ethanol. Deionized water was added for hydrolysis reaction. The sol solution was added slowly with a liquid solution of $NH_4OH_{(aq)}$ in ethanol till flurry-like precipitates appeared. In the deposition-precipitation route, tin chloride pentahydrate ($SnCl_4 \cdot 5H_2O$, Aldrich 98%) and tetrachloroauric acid ($HAuCl_4$, 99.9%) were used as the precursors. $HAuCl_4$ was hydrolyzed by $NaOH_{(aq)}$ followed by being dispersed in the aqueous suspension composed of tin oxide powders. For the two methods, washing with a water/ethanol solution was carried out till chlorine ions were not detected. The collected precipitates were dried at 30 °C for 48 h.

The obtained powders were characterized by X-ray diffractometer (XRD, Regaku, Miniflex II) equipped with Ni-filtered Cu-Kα radiation (15 mA, 30 kV and λ=1.5405 Å) and at a scan rate of 2°/min. Their surface composition and chemical state were examined with X-ray photoelectron spectroscopy (XPS, Model Quantera SXM, ULVAC-PHI, Kanagawa, Japan) with an Al-Kα radiation source (1486.6 eV) to excite photoelectrons in an ultra vacuum atmosphere (10^{-9} torr). The binding energy scale was calibrated by taking the C1s peak at 284.5 eV.

The gold/tin oxide powders were evaluated as sensing elements in a semiconductor carbon monoxide gas sensor. The sensor substrate used for this work is composed of a planar alumina substrate, a pair of gold electrodes, an oxide layer (sensing element) and a RuO_2 heater [7]. The oxide layer was fabricated by pasting oxide powders onto the Au-printed substrate followed by calcination in air. The sensor unit was placed inside a Pyrex glass chamber and evaluated in a static mode. Carbon monoxide was injected by a gas syringe into the test chamber and then immediately mixed with the background gas (air) inside the vessel. Humidity in the atmosphere was controlled at about 30-40% RH. The gas (air) flow for desorption process was set at 300 ml/min. A dual-channel DC power supply unit (Motech, Model PPS-1200) was used to maintain the circuit voltage at 4 V and provide the heater voltage. The sensor output voltage was read from a digital multimeter (Keithley, Model 2000). Application of a periodic heating voltage to the sensor is necessary in order to activate and clean the oxide surface.

RESULTS and DISCUSSION

Figure 1 shows XRD patterns of pure SnO_2 and Au-deposited SnO_2 powders prepared by co-precipitation (CP) and deposition-precipitation (DP) methods. These diffraction peaks were indexed to the tetragonal cassiterite structure of SnO_2 (JCPDS 21-1250). The peaks that are assigned to the metallic gold phases were not apparently observed. A possible explanation as to the cause is the small size of the gold particles deposited on the tin oxide surface.

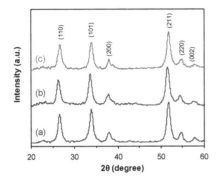

Figure 1. XRD spectra of powder samples (a) SnO_2 (500 °C calcined), (b) co-precipitation Au/SnO$_2$ (Au/Sn = 0.01, 500 °C calcined), and (c) Deposition-precipitation 3 wt% Au/SnO$_2$ (250 °C calcined).

Figure 2 shows the XPS spectra of the 3 wt% Au/SnO$_2$ powders prepared by the deposition-precipitation method and calcined at 250 °C in air for 2 h. The full scan spectrum revealed the presence of Sn (3d$_5$), O (1s) and Au (4f$_7$) elements on the particle surface (C 1s reference is taken at 284.5 eV). The Au doublet peak indicated that the gold species deposited on the oxide surface were characterized in the form of metallic Au^0 phases (with Au 4f$_{7/2}$ binding energy at about 84.1 eV [6]).

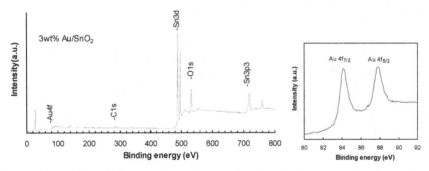

Figure 2. XPS full scan spectrum and Au 4f$_7$ spectrum of 3 wt% Au/SnO$_2$ powders prepared by deposition-precipitation method and calcined at 250 °C.

Figure 3 is a transmission electron microscopy (TEM) image taken for the 3 wt% Au/SnO$_2$ powders. The Au particle size is measured about 3.1±0.6 nm, and found to be located at the top of the tin oxide grain.

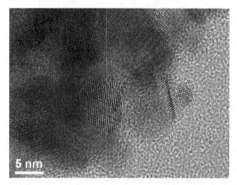

Figure 3. TEM image of 3 wt% Au/SnO$_2$ powders prepared by deposition-precipitation method.

Figure 4 shows the dynamic response curves of the SnO$_2$ and Au/SnO$_2$ powders exploited as sensing elements in a semiconductor sensor device for use to detect CO gas in air atmosphere at three concentrations (50, 250 and 500 ppm) and at 125 °C. Response is defined as the ratio of R_{air}/R_{co}, where R_{air} and R_{co} are the electrical resistances measured in the background air gas and CO gas, respectively. We observed the electrical resistance decreasing upon CO gas exposure (i.e., $R_{co} < R_{air}$). These transient response signals were quite reversible, with good baseline stability. Obviously, both Au/SnO$_2$ sensors (3 wt% Au/SnO$_2$ by DP method and Au/Sn=0.01 by CP method) exhibited superior performance than the SnO$_2$ sensor. This demonstrates the promoting effect of the gold particles deposited in the CO gas-sensing process. A possible explanation as to the cause is that the Au dopants would facilitate CO oxidation over the sensor surface, due to the formation of reactive oxygen ions O$_2^-$ on the gold surface [2].

$$O_{2\,(g)} + e^- \rightarrow O_2^-{}_{(ads)} \tag{1}$$

$$2CO_{(ads)} + O_2^-{}_{(ads)} \rightarrow 2CO_{2\,(ads)} + e^- \tag{2}$$

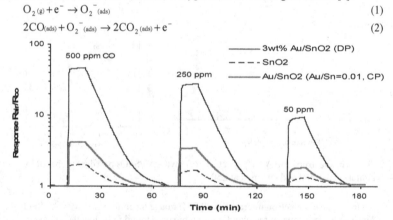

Figure 4. Dynamic response curves of SnO$_2$ and Au/SnO$_2$ powders based sensors to CO gas in air at three concentrations (50, 250 and 500 ppm) and at an operating temperature of 125 °C.

CONCLUSIONS

The present work completes the preparation of gold-deposited tin oxide nanocomposite powders and demonstrates the promoting effect of the gold particles to enhance sensing response to CO gas in air. These Au/SnO_2 composite powders are promising for practical applications in semiconductor sensors as well as catalytically active materials.

ACKNOWLEDGEMENTS

Financial support is gratefully acknowledged from the National Science Council of Taiwan under Grant Number NSC 100-2221-E-224-057.

REFERENCES

1. M. Ivanovskaya, P. Bogdanov, G. Faglia, G. Sberveglieri, *Sens. Actuators B: Chem.* 68, 344-350 (2000).
2. D. Buso, M. Post, C. Cantalini, P. Mulvaney, A. Martucci, *Adv. Funct. Mater.* 18, 3843-3849 (2008).
3. G. Neri, A. Bonavita, C. Milone, S. Galvagno, *Sens. Actuators B: Chem.* 93, 402-408 (2003).
4. M. Haruta, M. Daté, *Appl. Catal. A: Gen.* 222, 427-437 (2001).
5. M. Okumura, Y. Kitagawa, M. Haruta, K. Yamaguchi, *Appl. Catal. A: Gen.* 291, 37-44 (2005).
6. M. Khoudiakov, M.C. Gupta, S. Deevi, *Appl. Catal. A: Gen.* 291, 151-161 (2005).
7. C.-T. Wang, M.-T. Chen, *Sens. Actuators B: Chem.* 150, 360-366 (2010).

Mater. Res. Soc. Symp. Proc. Vol. 1449 © 2012 Materials Research Society
DOI: 10.1557/opl.2012.1171

A Study of Anodization Time and Voltage Effect on the Fabrication of Self- Ordered Nano Porous Aluminum Oxide Films: A Gas Sensor Application

Ildeman Abrego, Alfredo Campos, Gricelda Bethancourt and E. Ching-Prado[*].
Department of Natural Science, Faculty of Science and Technology, Technologic University of Panama, Republic of Panama

*Corresponding Author e mail: eleicer.ching@utp.ac.pa

ABSTRACT

The effect of time and voltage on the fabrication of self ordered nano porous aluminum oxide structure is studied. A two-step anodization process in 0.4M sulphuric acid and at low temperature was used to prepare structure with pore diameter between 20 to 29 nm. The surface morphology, including porous structure mainly, is characterized using scanning electron microscopy. Pore density 10^{10} to 10^{11} cm^{-2} is found. Thickness and electrical response are performed to these films. Capacitive sensor measurements for different water concentration are presented and discussed. The gas sensibility dependences with pore diameter and thickness layer are also reported.

INTRODUCTION

Anodized aluminum porous oxide films, have been intensively studied over the last five decades (1). This oxide presents a typical self- ordered nanochannel material (2,3). One of its more favorable applications is as template for the fabrication of structured materials with different technological applications (4,5). Another film feature is as gas sensor for humidity, ammonia and organic vapors (6,9). Also, noble metal electrodeposition in its pore that can bound to organic groups, such as –SH, -NC (10), or synthesizing carbon nanotube in its pore (11) has been done providing a new way for gas sensor fabrication.

The study of the variables that affect the preparation of self-organized, highly ordered aluminum oxide films with nano pores has attracted great interest in recent years (12). Aluminum oxide is prepared easily and at low cost by anodizing aluminum in different acids o acid mixtures, at varying temperature, voltage and time. This process produces a thin aluminum oxide barrier layer, over laid by ordered hexagonal alumina structure with a pore in its center (2).

The regularity of the porous structure is limited by the grain domain, but pore dimensions and structures features can be control by heat treatment and electro polishing before anodizing and by two or more anodization step with chemical etching in between (13). But, Masuda et al., asserted that the two-step anodization process appears to be the best choice as it results in high control of the nanostructure of Anodized Aluminum Oxide(AAO), such as its regularity and size(3).

Thus, the oxide film formed presents a high porous density (10^9 to 10^{12} cm^{-2}.)(14), thermal and physical stability (15) and highly ordered structure, promising features for a gas sensor device.

In this paper, we report on the effect of anodization time and voltage on film growth, pore distribution, density and size during the fabrication aluminum oxide nanostructure to be used as gas sensor.

EXPERIMENTAL

Sample preparation

High purity aluminum sample (99.95%) with one square centimeter (1.0cm^2) and 0.9mm thickness was used in the study. Before anodizing, the samples were degreased in acetone for 20 min, rinsed in de-ionized water and finally air dry. Next, it was annealed, under nitrogen ambient, at 500°C for 5 hours. To smooth the surface morphology, the sample were electropolished in a mixture H_2SO_4, H_3PO_4 and H_2CrO_4 at (75.0±2.0)° C for several minutes under constant volts.

Anodization was carried out in a conventional cell using a platinum electrode in parallel to the Aluminum anode, held a distance of 25 mm away. The cell was kept in an ice bath at (0 °C) and the electrolytic solution was stirred continuously during the anodization process, in order to maintain temperature and compositions of the 0.40M H_2SO_4 electrolyte homogeneous. To observe the effect of voltage on pore and cell structures, 16, 21, 26 and 31 volts were applied to the samples at determined constant time. To study the effects of the first and second anodized time on pore and cell structures at determined constant voltage, 60, 120, 180, 240 and 300 minutes were used.

After the first anodization step, the sample were cleaned in acetone and etched in a mixture of 6% w/w phosphoric acid (H_3PO_4) and 1.8% w/w chromic acid (H_2CrO_4) at (84.0±2.0)° C for 60 min to remove the porous oxide layer. Then, the sample was anodized again at identical condition as the first anodizing step. For each step, the variation of the anodizing current was recorded against time.

Sample characterization

After anodizing, the structure and morphology of the samples were examined by scanning electron microscopy (SEM, ZEISS EVO 40 VP). A top layer of sputtered conductive gold was deposited on the sample before electron microscopy was done. To these same samples, electrical conductivity was performed by placing a silver paint contact on the aluminum oxide layer and another on the pure aluminum back side. Impedance spectroscopy was done by placing the samples inside a glass chamber at room temperature and 20% relative humidity, using an impedance bridge (Agilent E4980A). Impedance spectra were recorded in the range of 20Hz to 2MHz. For Sensor measurements, the same samples were placed into a glass chamber connected to a heated glass reservoir where different volume of de- ionized water was injected with a micropipette and vaporized. Water vapor was carried on into the glass chamber by passing dry air using an air compressor. The air flux was adjusted to 4L/min. Sensor dc capacitance measurements were done by using a multimeter (Radio Shack Digital Multimeter) with RS232 interface at (23.0±2.0)°C temperature. Thickness film measurements were done with a Byko-test 4500 Coating Thickness Gauge.

DISCUSSION

Figure 1 show the currents transient recorded for first and seconds step anodic oxidation process for 26V and 120 minutes. During the first anodizing, there is a region associated with the first stage of pores growth in which the barrier layer is formed. This can be observed as a constant current at 250 seconds. This barrier layer facilitates the growth of the porous film, as can be seen in the second anodized where a constant current is acquired in less than 100 seconds.

This is attributed to the fact that there is already a way for the formation of pores. This same behavior is observed in all samples studied.

Figure 2a illustrates a typical SEM image of AAO film after a second step anodic oxidation, with cell expansion inset. The structure shows a hexagonal arrangement in ordered domains. In Figure 2b, the histogram shows a Gaussian distribution of pores diameters around (24.0±2.0) nm ??for 16V and 120 minutes of anodizing. For all samples, there is a similar Gaussian distribution and pore domains are also aligned with different degrees of order.

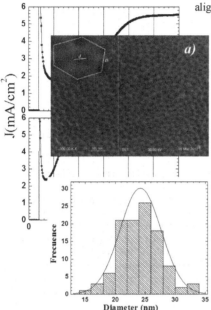

From the histograms, different average values are obtained ??for both, the diameter (*d*) and the distance between pores (*D*), as a function of voltage and anodizing time.

Figure 3 shows pore diameter average and distance between average pores in function of anodizing time and voltage. Figure 3a exhibits no significant variation in pore diameter and distance between them as a function of anodized time. Instead, in Figure 3b is observed that pore diameter increases with the applied voltage with a slope of 0.6 nm/V and pore distance between them increases with the applied voltage, with a slope of 0.9 nm/V. This is consistent with what is reported in the literature, that is, the AAO pore diameter depends only on electrolyte chemistry and anodizing voltage and not on the anodized time (16, 17). Pore density (# pore cm^{-2}) results show no significant variation with anodizing time. Instead for the voltage, the pore density changes from 10^{11}cm^{-2} for 16 volts to 10^{10}cm^{-2} for 31 volts.

Film thickness for AAO is in the order of 18μm to 43μm for 60 and 300 minutes anodizing time respectively. The results shows a linear relationship between AAO films thickness and anodizing time, with a rate of 0.14 μm/min for sample prepared at 0°C in 0.4 M sulfuric acid at 21.0 volt.

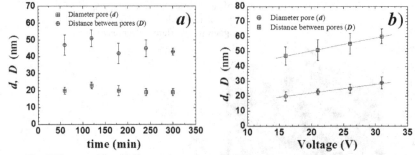

Fig. 3 Diameter (*d*) and distance between pores (*D*) versus anodized time and voltage

Figure 4 present the electric response capacitance, as function of time for different water volume content in the test sensor chamber. The figure shows bands associated to the sensitivity of water molecules. In these bands can be identified the water adsorption process (increasing capacitance) and desorption process (decreasing capacitance), respectively. It can be observed increasing in electrical response variation with increasing water volume. From figure 4 is possible to obtain the Relative Capacitance in function of water vapor volume. Figure 5 shows Relative Capacitance (?C/C) as a function of water vapor volume for sample anodized at two different voltage and time. Voltage effect on the Relative Capacitance is observed in figure 5a. For both sample, a well-behaved change in capacitance over the 20μL to 100μL range is observed. From the results presented in Figure 3, pore diameter is directly proportional to anodized voltage, thus pores surface for 21 volt sample is larger than for 16 volts. As pores diameter increase, many water molecules are adsorbed and diffuse into the pore side walls forming a thin water layer facilitating ions mobility, thus increasing charge accumulation. So water vapor sensitivity is expected to increase. In figure 5b, Relative Capacitance versus water vapor volume for two different anodizing times, shows an increase of the Relative Capacitance with time. It has been determined a direct proportion between film thicknesses with time which is expected to increase the total surface area of pore side wall. So, it is can expected an increment in ion mobility as well.

Figure 6 shows a Cole-Cole plot of the sample anodized at 31 V and 120 minutes. In the spectrum two contributions can be observed. One of them, at low frequencies (less than 10 kHz) seems to be characteristic of the electrical contacts. The other contribution is associated to membrane electrical properties and is evidenced at high frequencies (more than 10 kHz). The

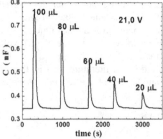

spectrum was interpreted according to a circuit equivalent model, where the membrane is fitted trough a simple R_mC_m parallel and the electrical contacts with parallel of the series Rct, W and Cdl (18).

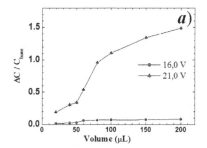 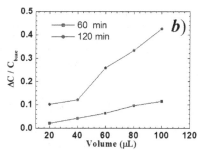

CONCLUSION

Aluminum oxides were prepared by anodization technique in order to fabricate self orders nanoporous structure. The results indicate that pores diameter (around 20 nm) and distance between pore (around 45nm) almost not change with the anodizations time. However they are dependent of the anodization voltage, because linear relationships were found. Also the study

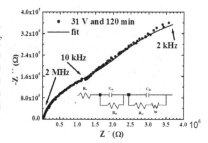

reveals that the pore density change from 10^{11} to 10^{10} cm^2 as the voltage increase. The electric capacitance measurements indicate that AAO sensitive to water vapor concentrations. This sensitivity increase with increasing anodizations voltage or with increasing anodization time.

ACKNOWLEDGMENTS

The authors are grateful to: SENACYT (projects FID 05-061 and APY-GC-10-046 A) for financial support; Smithsonian Tropical Research Institute and in special to Jorge Ceballos who helped us measuring, analyzing and discussing the morphological characterization (SEM); Institute of Materials Jean Rouxel- CNRS, France, for their assistance in AAO preparation methodology and Biomuseum, Amador Foundation, help us in film thickness measurements.

REFERENCES

[1] F. Keller, M.S. Hunter, D.L. Robinson, Structural features of oxide coatings on aluminum, J. Electrochem. Soc. 100 (1953) 411–419.

[2] O. Jessensky, F. Muller, U. Gosele, Self-organized formation of hexagonal pore structures in anodic alumina, J. Electrochem. Soc. 145 (1998) 3735–3740.

[3] H. Masuda, F. Hasegwa, S. Ono, Self-ordering of cell arrangement of anodic porous alumina formed in sulfuric acid solution, J. Electrochem. Soc. 144 (1997) L127–130.

[4] X. Duan, Y. Huang, Y. Cui, J. Wang, C.M. Lieber, Indium phosphide nanowires as building blocks for nanoscale electronic and optoelectronic devices, Nature 409 (2001) 66–69.

[5] T.M. Whitney, J.S. Jiang, P.C. Searson, C.L. Chien, Fabrication and magnetic properties of arrays of metallic nanowires, Science 261 (1993)1316–1319.

[6] RK Nahar, VK Khanna y W.Khole, On the origin of the humidity – sensitive electrical properties of porous aluminum oxide, J. of Physics. D. Applied Physics,17(1984)

[7] O.K. Varghese, D. Gong, W.R. Dreschel, K.G. Ong, C.A. Grimes, Amonia detection using nanoporous alumina resistive and surface acoustic wave sensors, Sensors and Actuators B 94(2003).

[8] Lujun Yao, Maojun Zheng, Haibin Li, Li Ma andWenzhong Shen, High-performance humidity sensors based on high-field anodized porous alumina films, Nanotechnology 20 (2009)

[9] Youngdeuk Kima, Bongbu Junga, Hunkee Leea, Hyejin Kimb, Kunhong Leeb, Hyunchul Parka, Capacitive humidity sensor design based on anodic aluminum oxide, Sensors and Actuators B 141 (2009

[10] Jong Bae Park,Youngsic Kim, Seong Kyu Kim and Haesseong Lee,Formation of Self-assembled Nanostructure on Noble Metal Islands Based on Anodized Aluminum Oxide,Bull.Korean Chem. Soc.,Vol. 25, No. 4(2004)

[11] Hyun Tae Chun, Dong gu Lee, You Suk Cho, Se Young Jeong, Dojin Kim, A Study on Gas Sensor Based on Carbon Nanotubes on Anodized Aluminum Oxide, Mol. Cryst. Liq. Cryst., Vol. 459, pp. 231(2006)

[12] Rajvinder S. Virk, Study of voltage, acid concentration, and temperature on nanopore structures, Master's Theses, San Jose State University(2008)

[13]A.P. Li, F.Muller, A. Birner, K. Nielsch and U. Gösele, Polycrystalline nanopore arrays with hexagonal ordering on aluminum, American Vacuum Society[SO734-2101(99)20304-1](1999)

[14] Son H.Le, Angela Camerlingo, Hoa T.M. Pham2, Behzad Rejaei1, Pasqualina M. Sarro, Anodic Aluminum Oxide Templates for Nanocapacitor Array Fabrication, Delft University of Technology, P.O Box 5053, 2600 GB, Delft, the Netherlands

[15] Traversa E 1995 Sensors Actuators B 23 135.

[16] F. Li, L. Zhang, R. Metzger, On the growth of highly ordered pores in anodized aluminum oxide, Chem. Mater. 10 (1998) 2470-2480.

[17] H. Masuda, K Fukuda, Ordered metal nanohole arrays made by a two-step replication of honeycomb structure of anodic alumina Science 268 (1995), 1466-1468

[18] P. Bocchetta, R. Ferrano, F. Di Quarto, Advance in anodic alumina membranes thin film fuel cell: JPS, 187 (2009) 49-56.

Mater. Res. Soc. Symp. Proc. Vol. 1449 © 2012 Materials Research Society
DOI: 10.1557/opl.2012.792

PbS Nanoparticles: Synthesis, Supercritical Fluid Deposition, and Optical Studies

Joanna S. Wang[a], Bruno Ullrich[a,b], and Gail J. Brown[a]

[a]Air Force Research Laboratory, Materials & Manufacturing Directorate, Wright Patterson AFB, OH 45433-7707, USA

[b]Instituto de Ciencias Físicas, Universidad Nacional Autónoma de México, Cuernavaca, Morelos, México C.P. 62210

ABSTRACT

Lead sulfide (PbS) nanoparticles (NPs) of different sizes (2.0 nm - 14.4 nm) have been synthesized in our laboratory. By using those NPs, we formed colloidal films on glass and GaAs substrates employing a specialized supercritical fluid CO_2 (sc-CO_2) deposition method. The deposited films contain only the PbS NPs and the protecting group of oleic acids and require no polymer matrix. The NP films are solvent free, environmentally stable, and show good adhesion to the substrates. The sc-CO_2 deposition process can deposit films ranging in thickness from a few monolayers, in well ordered arrays, up to 0.5 μm or greater. The photoluminescence (PL) properties of these nano-structured films were studied with Fourier transformation infrared spectroscopy from 5 K up to 300 K.

INTRODUCTION

The intrinsic features of PbS nanoparticles, particularly the clearly enhanced quantum confinement with respect to other attractive semiconductors such as CdTe and CdS, has boosted the research on PbS NPs during the past 15 years [1-5]. Particularly, the emission properties have attracted considerable attention [5-9]. While many of the PbS colloidal NPs studied have been embedded in glass or polymer matrices, we sought to develop a deposition process that could be compatible with standard semiconductor device processing, i.e. a cleaner process with less solvent residue and a process that enables close contact between the semiconductor surface and the NPs. We employed supercritical fluid CO_2 processes due to its near zero surface tension, the ability for the removal of solvents, and the capability to arrange NPs in ordered arrays [10-12] - a task difficult to achieve by traditional solvent deposition. To study the interactions of PbS NPs with semiconductor substrates colloidal PbS films of different NP diameters are highly useful. Glass substrates were used as control samples to isolate substrate charge transfer effects. For this study we developed both PbS NP synthesis and deposition processes using supercritical fluid CO_2. PbS nanoparticles precipitate evenly and self-assemble to form a uniform 2-D array on TEM copper grids, glass, and GaAs substrates during the sc-CO_2 deposition.

EXPERIMENTAL DETAILS

PbS QDs of different sizes have been selectively synthesized by the alteration of the following parameters: oleic acid/octadecene ratios, injection temperature, and growth temperature [13] during the synthesis. The TEM images of various PbS NPs synthesized with this process, ranging from 2.0 nm to 14.4 nm in diameter, are shown in Figure 1 below.

Figure 1. TEM images of PbS QDs with different sizes.
a: size = 14.4±1.6 nm, b: size = 8.6±1.1 nm,
c: size = 4.8±0.54 nm, d: size = 2.8±0.31 nm

Using the supercritical fluid CO_2 high pressure system (Fig. 2a) with our homemade sample deposition apparatus (Fig. 2b), we generated thin films and close-packed arrays. Samples were fabricated on a glass substrate to avoid any electronic charge transfer between the NPs and the inert insulating substrate. We developed a special process for creating uniform NP films over a 1 cm diameter area on glass slides using the apparatus in Fig. 2b.

The sample deposition apparatus consists of two aluminum plates, each with 2 circular 1 cm opening in the center. The specific design enables the production of two samples by deposition circle, which increased the coincident characterization activities and sample output. A piece of glass was inserted between the plates with two Teflon o-rings placed on each side of the glass substrate in order to prevent leaking of the QD solution. 40 μL of PbS solution (depending on the thickness of PbS film required) at concentration of 10 μg/μL were spiked into 200 μL toluene solvent, followed by transferring the solution prepared into the sample holder in Fig. 2b. Using this adapted deposition device, the particular chemistry of the sc-CO_2 process was studied and optimized by feedback from the optical characterization.

The sc-CO_2 deposition process was carried out using a 35.3 mL high-pressure stainless steel chamber. The chamber was charged with liquid CO_2 (60 atm) at room temperature over a period of 10 min and then the pressure was raised to 80 atm. The system was then heated to 40 °C to convert the liquid carbon dioxide to the supercritical fluid phase. At this time the pressure inside the chamber was about 145 atm. The ISCO pump then slowly raised the pressure up to 180 atm in the chamber. The high pressure chamber was left at this condition (40 °C and 180 atm) for 30 min to ensure the system pressure is consistent and reproducible. In this process the PbS NPs are precipitated from the toluene solution by a gas-antisolvent (GAS) mechanism described previously in the literature [11,14], where an increasing amount of CO_2 alters the polarity of the toluene solvent and it becomes unfavorable for particle stabilization in the colloid, thus resulting in the particles precipitation from solution. Not only are the NPs uniformly deposited on the substrate in a dense film, but also the toluene is completely removed by the sc-CO_2 cleaning process [14,15]. Fig. 2c shows a uniformly formed PbS sample with an areal density of 0.41 mg/cm^2 and a thickness of 0.854 μm.

A different deposition process was required for the more fragile GaAs substrates. In this case, small pieces of heavily doped (5.2×10^{18}-1.4×10^{19} cm^{-3}) p-type GaAs:Zn (0.5 x 0.5 cm^2) were immersed in a metal sulfide nanoparticle/toluene solution placed in a small vial. A schematic diagram of this process is shown in Fig. 2d. Similar sc-CO_2 process conditions as described above were used to precipitate the PbS nanoparticles evenly on the p-GaAs and remove the solvent from the vial. The resulting film on p-GaAs is shown as a SEM image in Fig. 2e. Unlike a benchtop solvent evaporation process, which has a high surface tension at the liquid/vapor interface and leads to imperfect nanoparticle ordering forming isolated islands, percolating domains, locally high particle population and uneven surface coverage [15-17], the sc-CO_2 process which has near zero surface tension leads to very uniform and densely packed films.

For the PL experiments, the sample was excited with the continuous wave (cw) emission at 532 nm of a solid state laser by keeping the impinging intensity below 50 W/cm^2. The PL emission was detected with a nitrogen cooled InGaAs detector attached to a BOMEM DA3 FTIR spectrometer. In order to measure the PL spectra at cryogenic temperatures, the sample was mounted in a closed cycle optical helium gas cryostat operating in the range 5 K- 300 K.

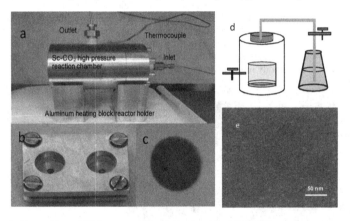

Figure 2. Deposition of PbS nanoparticles in supercritical fluid CO_2.
(a) supercritical fluid CO_2 high pressure system, (b) home-made sampling apparatus and (c) deposited film using supercritical fluid CO_2. (d) schematic diagram of PbS deposition on GaAs substrate in sc-CO_2 and (e) SEM image of PbS on GaAs substrate using sc-CO_2 deposition method.

DISCUSSION

The traditional way to synthesize nanoparticles in our lab was using a water-in-oil based microemulsion method [10-12]. However, the microemulsion technique did not work for synthesis of PbS nanoparticles according to our experiments. The reason is likely due to the fact that the aqueous chemistry of lead ions is complicated and various lead species can exist in water depending on pH value and redox environment. It is difficult to control the dissolved lead ions in the aqueous solution to be in the pure Pb^{2+} state. Therefore, we chose to fabricate PbS nanoparticles by a traditional synthesis in organic solvents with oleic acid as a capping agent [13].

Figure 1 above shows TEM images of PbS with different sizes. PbS QDs synthesized as 14.4 nm are cubic lattice structures with tendency of self assembly showing a very ordered array (Fig. 1a). The 14.4 nm (Fig. 1a) and 8.6 nm (Fig. 1b) NPs have no absorbance features in near infrared region. The smaller PbS NPs, 4.8 nm (Fig. 1c) and 2.8 nm (Fig. 1d), show strong absorbance in UV-Vis-NIR region, and strong PL intensities with quantum confinement effects.

Optical microscope image results (Figures 3a, 3b, 3c and 3d) demonstrate the surface coverage features of the films on glass using both solvent and sc-CO_2 deposition methods. Apparently, the sc-CO_2 deposition method provides a more featureless surface with respect to the solvent deposition method. Optical images generated from sc-CO_2 (Figures 3b and 3d) demonstrate uniform and even coverage, whereas from solvent deposition method, coffee cup ring structures were shown in both Figures 3a and 3c.

Our XRD spectra indicate the PbS films have extended stability and PbS XRD pattern of PbS nanoparticles agrees very well with the standard reference data (JCPDS) for PbS galena structure, after nearly 2 year period of shelf life time. The film thickness was measured with DekTak 6 M Stylus Profiler instrument. 3000 μm scan length was used for a standard scan. The average scanned height corresponds to the PbS film thickness, for instance, the film thickness in Figure 2c, is 0.854 μm. The films on the glass deposited by sc-CO_2 are reasonably adhered to the glass surface and are not readily removed by an adhesive tape.

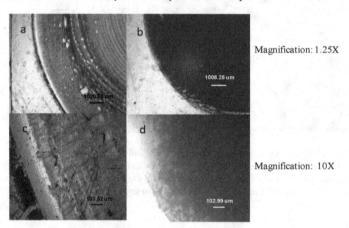

Magnification: 1.25X

Magnification: 10X

Figure 3. Microscope images of PbS films deposited on glass substrates. (a) solvent deposition and (b) sc-CO_2 deposition at magnification of 1.25X. (c) solvent deposition and (d) sc-CO_2 deposition at magnification of 10X.

For the PL study, PbS quantum dots (QDs) with diameters of 4.7 nm and 2.0 nm were used, motivated by the room temperature emission wavelengths in the range 1.0 μm-1.55 μm, which is important for telecommunication applications. Indeed, the absorbance peak of the PbS solution of 4.7 nm QDs used to form the films took place at around 1330 nm (0.932 eV) at 300 K. The optical properties of the PbS nanoparticle (4.7 nm diameter) films were measured by PL spectroscopy. The samples fabricated by the sc-CO_2 deposition process showed clearly detectable room temperature emission at 0.84 eV (1470 nm), which is blue-shifted from the bulk value at 0.41 eV due to quantum confinement effect. The PL wavelength of PbS QDs in solid films, versus in solution, show a red shift with increasing concentrations.

Figure 4 shows the emission of the PbS/p-GaAs sample at 5 K and 300 K. The spectra show the pronounced PbS QD emission and the photoluminescence of the p-GaAs at a low temperature (5 K). The intensity of PL emission at 5 K is about 3 times higher than that at room temperature (300 K). PbS films on glass and a GaAs substrate, formed by the sc-CO_2 deposition method, show a strong photoluminescence intensity attributed to the narrow particle size distribution and homogenous particle morphology, while PbS films deposited by traditional solvent deposition method have weak signal intensities attributed to particle non-uniformity.

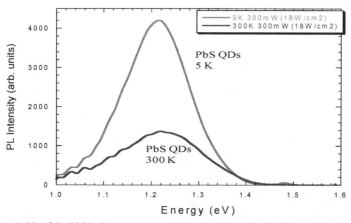

Figure 4. PL of the PbS/p-GaAs sample at 5 K and 300 K excited with a laser intensity of 18 W/cm^2. The emission of the GaAs substrate was observed in addition to the PL of the QDs.

CONCLUSION

PbS QDs in different sizes were synthesized in our laboratory. An apparatus for depositing PbS QDs on glass substrates employing a sc-CO_2 solution deposition method is described. The PbS QDs deposited in this way have shown improved uniformity and a good coverage in comparison with conventional solution deposition. PL emission of the PbS/p-GaAs sample at 5 K and 300 K were measured in this study. The spectra show the pronounced PbS QD emission at a low temperature (5 K). The intensity of PL emission at 5 K is about 3 times higher than that at room temperature (300 K). PbS films on glass and GaAs substrate formed by the sc-CO_2 deposition method show strong PL intensity attributed to the narrow particle size distribution and homogenous particle morphology. An important future application of PbS quantum dots will be the formation of optoelectronic hybrid devices. Prospective partner for these hybrids are the III-V compound GaAs and related semiconductors, which are the current main players in the field of optoelectronic devices.

REFERENCES

1. L. Bakueva, S. Musikhin, M. A. Hines, T. W. F. Chang, M. Tzolov, G. D. Scholes and E. H. Sargent, *Appl. Phys. Lett.* **82(17)**, 2895 (2003).
2. J. J. Peterson and T. D. Krauss, *Nano Lett,* **6(3)**, 510 (2006).
3. A. Osherov, J. P. Makai, J. Balazs, Z. J. Horvath, N. Gutman, A. Sa'sr and Y. Golan, *J. Phys.: Condens. Matter* **22**, 262002, (2010).
4. R. D. Schaller, J. M. Pietryga, S. V. Goupalov, M. A. Petruska, S. A. Ivanov and V. Klimov, *Phys. Rev. Lett.* **95**, 196401 (2005).
5. R. Dalven, *Infrared Phys,* **9**, 141, 1969.
6. V. Biju, R. Kanemoto, Y. Matsumoto, S. Ishii, S. Nakanishi, T. Itoh, Y. Baba and M. Ishikawa, *J. Phys. Chem.* C. **222**, 7924, (2007).
7. T. Y. Liu, M. Li, J. Ouyang, M. B. Zaman, R. Wang, X. Wu, C. S. Yeh, Q. Lin, B. Yang and Kui Yu, *J. Phys. Chem.* C. **113**, 2301, (2009).
8. K. A. Abel, J. Shan, J. C. Boyer, F. Harris and F. C. J. M. Veggel, *Chem. Mater.* **20**, 3794 (2008).
9. T. Zhang, H. Zhao, D. Riabinina, M. Chaker and D. Ma, *J. Phys. Chem.* C. **114**, 10153, (2010).
10. J. S. Wang, A.B. Smetana, J. J. Boeckl, G. J. Brown and M. Wai, *Langmuir* **26(2)**, 1117 (2010).
11. A. B. Smetana, J. S. Wang, J. J. Boeckl, G.J. Brown and C. M. Wai, *J. Phys. Chem. C* **112**, 2294 (2008).
12. A. B. Smetana, J. S. Wang, J. J. Boeckl, G. J. Brown, C. M. Wai, *Langmuir* **23**, 10429 (2007).
13. M. A. Hines, and G.D. Scholes, *Adv. Mater.* **15**, 1844 (2003).
14. J. Liu, M. Anand and C. B. Roberts, *Langmuir* **22**, 3964 (2006).
15. M. C. McLeod, C. L. Kitchens, C. B. Roberts, *Langmuir* , **21**, 2414, (2005).
16. X. M. Lin, H. M. Jaeger, C. M. Sorensen and K.J. Klabunde, *J. Phys. Chem. B* **105**, 3353 (2001).
17. P. C. Ohara, and W. M. Gelbart, *Langmuir* **14**, 3418 (1998).

Mater. Res. Soc. Symp. Proc. Vol. 1449 © 2012 Materials Research Society
DOI: 10.1557/opl.2012.1288

Controlled Synthesis of Si Nanopillar Arrays for Photovoltaic and Plasmonic Applications

Umesh Gautam[1], Jun Wang[1], Dilip Dachhepati[1], Seyyedsadegh Mottaghian[1], Khadijeh Bayat[2] and Mahdi Farrokh Baroughi[1]
[1]Department of Electrical Engineering and Computer Science, South Dakota State University, Brookings, SD 57007
[2]Harvard School of Engineering and Applied Sciences, 29 Oxford Street, Cambridge, MA 02138, USA

ABSTRACT

This paper focuses on developing a robust process to independently control the geometrical parameters of Si nano-pillar (NP) arrays. These parameters include height and diameter of NPs, spacing between them, and the shape of the NPs. We have shown that the diameter, height, and spacing of NPs can be independently engineered by controlling the diameter of nano-beads through synthesis procedure, duration of isotropic SiO_2 etching and duration of anisotropic Si etching, respectively.

INTRODUCTION

Nanostructure arrays have shown strong potential for enhancing optical path length and light scattering in photovoltaics (PVs) and near field electromagnetic field enhancement for plasmonic applications [1]. Nano-pillar (NP) arrays have found promising applications in the fields of PVs and plasmonics. In the field of PVs, nanostructure have been utilized to 1) reduce reflection losses at the surface of solar cells and 2) break the trade-off between light absorption and charge collection by replacing vertical charge transport pathways with a radial path [2,3]. Independence of light absorption from charge separation in these solar cells enables them to achieve a reasonable PV performance even on very low quality, in terms of excess carrier lifetime, PV materials [4].

Significant progress has been made for development of methodologies for formation of Si NP in recent years [5]. Si NPs have been synthesized by chemical vapor deposition (CVD) via vapor-liquid-solid (VLS) growth technique. Kendrick et al. reported radial junction Si solar cells using Si wire arrays grown by Au catalyzed VLS growth on patterned Si substrates and achieved a power conversion efficiency of 2.3% [6]. The remnants of metal atoms from the catalyst in the Si wires behaved as deep traps and seriously reduced the open circuit voltage of the device. Optical and electron beam lithography have been utilized for development of Si NP arrays [7] However, optical and electron beam lithography in deep sub-micron regime are expensive processes. Self-assembly of silica nano-beads followed by reactive ion etching of Si substrate using the self-assembled particles as the etching mask, has gained significant attention for development of Si NPs. This method, so called nano-sphere lithography, was utilized by E. Garnett et al. for development of vertically aligned Si NPs over a large area of 4x4 cm^2 [1].

Successful utilization of this method in PVs and plasmonics require stringent control on the geometrical parameters of the NP array [4, 8-10]. For example, Paudel et al. showed that field enhancement in 2.5 dimensional plasmonic crystals is highly dependent on the height, spacing, and diameter of NPs [Paudel 2009]. It is, therefore, very important to design a robust fabrication scheme capable of independent control over the geometrical parameters of NPs. This paper focuses on developing a robust process to independently control the height and diameter of NPs, spacing between them, and the shape of the NPs.

EXPERIMENT

Nano-sphere lithography relies on silica nano-beads with a dimension comparable to that of the diameter of NP. A popular synthesis process for silica nano-beads is hydrolysis of tetraethyl orthosilicate (TEOS) in solution containing ethanol, water, and ammonium hydroxide [11, 12]. The size and distribution of the nano-beads highly depends on the reactant concentration, time and the reaction temperature.

These nano-beads can be coated on the Si substrate by dip coating method that leads to self-assembly of nano-beads on the substrate in compact lattice structure then be used as etching mask for fabrication of NP array. The quality of the self-assembled monolayer highly depends on the concentration of the silica nano-beads in the aqueous solution, which is used for dip coating. The rate of evaporation also plays a key role in determining if the mask layer forms a monolayer or a multilayer. The spacing between the nano-beads can be tuned by isotropic dry etching of the nano-beads to a desired extent. The height of the NPs is determined by the duration of the deep trench etch. Therefore a proper aspect ratio and suitable spacing between the NPs can be achieved by proper selection of the proportions of the reactants (to achieve desired diameter), oxide etching time (to determine spacing between the pillars) and the vertical etching that determines the height of the pillars.

60 ml of water, 60 ml of ammonium hydroxide 240 ml of ethanol and 10 ml of tetraethyl orthosilicate were stirred at 300 rpm for 12 hours at room temperature. The solution was then centrifuged at 3500 rpm and washed with deionized water. After three cycles of centrifuge and washing by water, the silica particles were mixed in 60 ml of deionized water. The Czochralski Si wafers were cleaned with Piranha at 120 °C for 30 minutes after sonicating in acetone and isopropanol, respectively. The Si substrates were dried with nitrogen gas after isorpropanol dipping. The clean substrates were then dipped into the solution and left to dry. With proper match of duration of dip and the evaporation rate a single layer of the nano-beads was achieved over a fairly large area. The silicon substrates coated with self assembled layer of SiO_2 were etched with STS etcher (reactive ion etcher deposition system) for various time intervals of 2 to 7 minutes. The process was carried out at pressure of 75 mT, RF power 150 watts, Ar flow rate 50 sccm, CF_4 flow rate 25 sccm and CHF_3 flow rate 50 sccm. After tuning the diameter so as to change the spacing between the NPs, deep trench etching was carried out for various time duration to get different aspect ratio of the NPs. The deep trench etching consisted of three steps with main etching step carried out at pressure 8 mT, RF1 power 10 watts, RF2 power 700 watts, C_4F_8 flow rate 63 sccm, SF_6 flow rate 27 sccm, Oxygen flow rate 10 sccm. After the etching processes the sample was cleaned with plasma asher for 5 minutes using oxygen flow rate of 100 sccm at 300 mTorr and 200 W RF power. In order to remove the silica nano-bead mask the sample was cleaned with HF: H_2O=1:100 (v/v) solution. The sample was finally treated with HF: HNO_3=1:99 (v/v) solution to smoothen the shape of the NPs.

To make the metallic NPs for plasmonic application, 1500 Å gold layer was deposited on top of this Si NP array with chromium adhesion layer of 150 Å using evaporation deposition method. The scanning electron microscope (SEM) and Color 3D laser scanning microscope imaging was done to study the morphology of the nano-mask and the NP array thus fabricated.

RESULTS AND DISCUSSION

Figure 1(a) shows the SEM image of the single layer of nano-beads dipcoated on the Si substrate. The nano-beads are arranged in an excellent short range, closely packed order; crystal directions are represented by arrows. Although good long range order with crystal regions with up to 200 μm x 200 μm area has been achieved, further improvement is necessary to obtain maximum surface coverage. Nevertheless, minimal distortion due to point defects (smaller or larger nano-beads) and linear defects can be observed. These small distortions add up in the long run and lead to crystal domains with random orientation with respect to a reference domain (Figure1(b)).

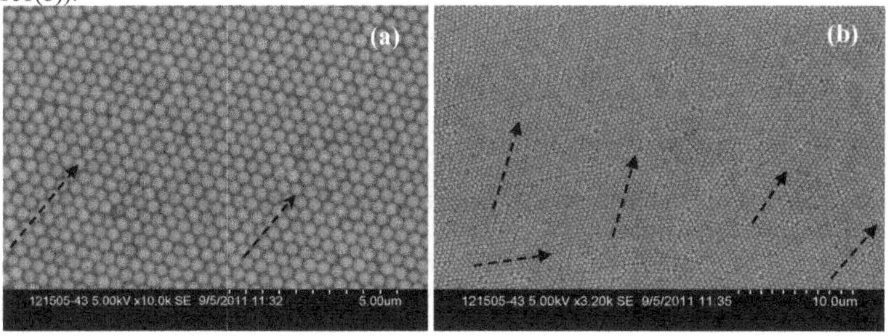

Figure 1: (a) SEM image of top view of self-assembled nano-beads (b) demonstration of crystal domains with random orientation

Very uniform distribution in terms of position and the size of the nano-beads, except for few of the interstitial vacancies, is visible. To further study the geometry of the nano-mask, MATLAB simulation was carried out in a section of the SEM image in Figure 1(a) for the coordinate analysis of the position of the nano-beads. It was possible to locate almost all the nano-beads in the SEM image by simulation except few sliced particles at the edge of the image as shown in Figure 2. Similar analysis was carried over the entire image to study the uniformity in the size of the nano-beads.

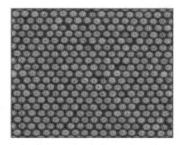

Figure 2: MATLAB analysis to study the position, spacing and bond angle of the nano-beads forming the nano-mask

The spacing between the adjacent nano-bead centers in the close-packed nano-mask is fairly uniform as demonstrated in Figure 3 (a). The distance between the centers of the adjacent nano-beads was 400nm for the majority of the nano-beads. Similarly the statistical analysis of bond angle for the close-packed structure is demonstrated in Figure 3 (b). For majority of the nano-beads forming the hexagon around a center nano-bead, the bond angle is 120°.

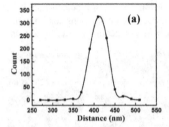 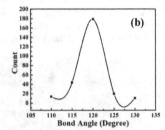

Figure 3: (a) Statistical analysis of the spacing between the nano-beads (b) Bond angle of hexagonal close-packed nano-mask

Similar analysis was performed on broader area of nano-mask shown in Figure 1 (a). The distribution of the size of these nano-beads demonstrated in Figure 4 verifies that the majority of them have uniform diameter of 360 nm. Although there are particles with diameter as large as 440 nm and as small as 320 nm, their count is negligibly small. This makes the nano-mask uniform in terms of distribution on the substrate and their diameter.

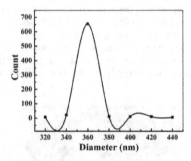

Figure 4: Size distribution of the SiO_2 nano-beads in Fig. 1(a)

The Si wafer with nano-mask on it was anisotropically etched by the deep trench etcher for various durations to get NPs with different heights. Figure 5 (a) shows SEM image of an array of Si NPs synthesized by deep trench etching through the silica mask. The NPs were cylindrical and standing upright with silica nano-beads on top of them. The Silica nano-beads were washed away with HF/H_2O solution. The NP array after HF/H_2O treatment is shown in Figure 5(b).

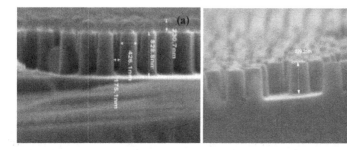

Figure 5 (a) NP array after deep trench etching through Silica mask (b) NP array after HF/H$_2$O treatment

The shape of the cylindrical Si NPs was further modified by wet chemical processes. By etching the NPs with a solution of HF and HNO$_3$ for very short time duration, the cylindrical shape was modified towards a nanocone structure as shown in Figure 6 (a).

In order to tune the spacing between the NP centers and the aspect ratio of the NPs, isotropic etching of the silica nano-mask was carried out for various time durations 2 to 5 minutes followed by the deep trench etching for 3 to 8 minutes. This method produced high aspect ratio NP arrays as shown in Figure 6 (b) for isotropic etching of 4 minutes and deep trench etching of 7 minutes.

Figure 6 (a) SEM image of NP array after wet chemical etching of the NPs (b) High aspect ratio NPs obtained by isotropic etching of nano-beads followed by reactive ion etching

The Si NP array synthesized using a self assembly and dry and wet etching techniques was used to develop a metallic grating for plasmonic application. Figure 7 shows cross sectional view of an array of gold NPs synthesized by evaporation deposition of gold on the Si NPs. The conformal deposition of gold on Si using the lift off fixture replicated the shape of Si NPs on gold. The dimensions of Si NPs were chosen such that gold NPs obtained had desired spacing and diameter even though gold deposits on the side-walls of the NPs and in the space between them. Thus this method was successfully implemented to fabricate plasmonic structures by self assembly as a low cost alternative to electron beam lithography.

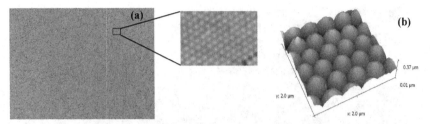

Figure 7: Cross sectional view of highly ordered self-assembled plasmonic crystal obtained by deposition of a gold layer on the surface of Si NP array over a fairly large area (a) Confocal laser microscope image and zoomed section of the image showing perfect periodicity (inset) (b) AFM image of a section of the metal NP structure.

CONCLUSIONS

The different geometric parameters of the self assembled NPs were independently controlled. The diameter was controlled by changing the reactant concentration to achieve a desired diameter of the nano-beads. The spacing between the NPs was tuned by isotropic etch of the nano-beads. The height of the NPs was determined by the duration of anisotropic etch. By the proper selection of the initial diameter of the nano-beads, isotropic etch time and the anisotropic deep trench etch time a NP array with desired aspect ratio and spacing was achieved. Gold nano-pillar array was fabricated by e-beam evaporation deposition of gold on this NP array.

ACKNOWLEDGMENTS

This work was supported by National Science Foundation-EPSCoR grant number 0054609 and the state of South Dakota. The authors would like to thank Nanofabrication Center of the University of Minnesota, a node of National Nanotechnology Infrastructure Network, for access to Deep Trench Etcher, STS Etcher and CHA E-Beam Evaporator facility.

REFERENCES

1. E. Garnett and P. Yang, Nano Lett. 10, 1082-1087 (2010)
2. Y. Lu and A. Lal., Nano Lett., 10, 465–4656 (2010)
3. L. Muskens, J. G. Rivas , R. E. Algra, M. Bakkers and A. Lagendijk, Nano Lett. 8, 2638 (2008)
4. B. M. Kayes and H. A. Atwater, N. S. Lewis, J. Appl. Phys. 97, 114302 (2005)
5. V. Schmidt, J. V. Wittemann and U. Gosele, Chem. Rev. 110, 361-388 (2010)
6. C.E. Kendrick, H.P. Yoon, Y.A. Yuwen, G.D. Baarber, H. Shen, T. E. Mallouk, E. C. Dickey, T. S. Mayer and J. M. Redwing, Appl. Phys. Lett. 97, 143108 (2010)
7. Y. K. Choi, J. Zhu, J. Grunes, J. Bokor and G. A. Somorjai, J. Phys. Chem. B 107, 3340-3343 (2003)
8. B. M. Kayes, M. A. Filler, M. C. Putnam, M. D. Kelzenberg, N. S. Lewis, and H. A. Atwater, Appl. Phys. Lett. 91, 103110 (2007)
9. H. P. Paudel, K. Bayat, M. F. Baroughi, S. May and D. W. Galipeau, Optics Express Vol. 17, No. 24, 22180 (2009)
10. H. P. Paudel, M.F. Baroughi and K. Bayat, J. Opt. Soc. Am. B /Vol. 27, No. 9 (2010)
11. W. Stober, A. Fink and E. Bohn, J. Coll. Interf. Sci. 26, 62 (1968)
12. G. Kolbe, Dissertation, Jena, Germany (1956)

Mater. Res. Soc. Symp. Proc. Vol. 1449 © 2012 Materials Research Society
DOI: 10.1557/opl.2012.1289

Solution Growth and Optical Characterization of Thin Films with $ZnO_{1-x}S_x$ and ZnO Nanorods in Core-Shell like Nanostructure for Solar Cell Application

Ratheesh R. Thankalekshmi and A. C. Rastogi

Department of Electrical and Computer Engineering and Center for Autonomous Solar Power (CASP), Binghamton University, State University of New York, Binghamton, NY, 13902, USA

ABSTRACT

ZnO films with a nanostructure dominated by 150-200 nm size highly c-axis oriented nanorod arrays were deposited by hydrothermal synthesis over surface activated quartz substrates. Sulfur infiltration and growth of $ZnO_{1-x}S_x$ over ZnO nanorods was carried out by chemiplating process using slow hydrolysis of thiourea solution at 95°C. Formation of $ZnO_{1-x}S_x$ nanocrystals of 20-30 nm size over (0001) facets of the ZnO rods is shown. With progressive growth of $ZnO_{1-x}S_x$ nanocrystal and full ZnO nanorod coverage, the formation $ZnO/ZnO_{1-x}S_x$ core –shell nanostructure is realized. X-ray photoelectron spectroscopy analysis shows chemical shifts in O1s and S2p spectra confirming the formation of $ZnO_{1-x}S_x$ ($0.1 \leq x \leq 0.2$) nanocrystal shell. Reduction in optical band gap from a 3.24 eV for ZnO nanorod core to 2.78 eV for the $ZnO_{1-x}S_x$ shell is consistent with the band gap bowing effect due to sulfur addition over the ZnO nanorod surface.

INTRODUCTION

Nanostructured zinc oxide with high band gap, large free-exciton binding energy (60meV) and high electrical conductivity in the doped form has attracted considerable attention due to potential application in the ultraviolet optoelectronics, field effect devices and sensors [1]. Recently, one dimensional (1-D) ZnO/ZnS core–shell nanostructures, and heterostructures are being actively investigated for application in multifunctional electronic devices [2] and in the polymer and dye based photovoltaic solar cells [3]. ZnO/ZnS nanostructure in solar cells facilitate band alignment thereby reduce the charge recombination rates and enhance carrier collection [4]. The most studied nanostructure is the ZnO nanowire arrays synthesized by vapor transport [5] and hydrothermal [6-9] methods and to form ZnO/ZnS core-shell structure these are over coated with ZnS. To tailor electrical and optical properties, anion doping of ZnO with sulfur to form zinc oxysulfide $ZnO_{1-x}S_x$ ($0 < x \leq 0.4$) alloyed film appears attractive. Growth of homogeneous $ZnO_{1-x}S_x$ thin films by oxidation of ZnS, reactive sputtering and atmospheric vapor transport techniques has been reported [10]. Nanostructured $ZnO_{1-x}S_x$ thin films have however received a limited attention [11], in particular, the $ZnO_{1-x}S_x$ shell structure over ZnO nanorods have not been studied. In solar cells, $ZnO_{1-x}S_x$ shell with engineered band gap can harvest more visible radiation and low valence band offset facilitate carrier transport across the junction in CIGS solar cells. We have synthesized nanocrystalline $ZnO_{1-x}S_x$ shell structure over ZnO nanorod arrays in aqueous medium at lower temperatures by surface conversion. This paper reports on the structural, compositional and optical properties of such nanostructures.

EXPERIMENT

ZnO films with nanorod arrays were deposited by hydrothermal synthesis over surface activated quartz substrates. Surface activation was done by depositing a thin ZnO film using the chemical spray pyrolysis of 0.1M zinc chloride solution dissolved in 70% isopropyl alcohol and 30% de-ionized water mixture. Typically the solution was air-sprayed at a ~30 cm distance, spray rate ~3 ml/min, atomizing spray nozzle pressure ~ 10 psi and substrate temperature ~300°C. Hydrothermal synthesis of ZnO nanorods was done in an aqueous solution of 0.03M zinc nitrate hexahydrate, 0.03M hexamethylenetetramine (HMT) and traces of thiourea (Tu) as an activator. The surface activated quartz substrates were kept vertically in the precursor solution in sealed glass bottles at ~95°C for ~ 8 hours. After completion of ZnO growth, the coated substrates were rinsed in de-ionized water to remove any residual precursor film and dried in air. For the synthesis of the $ZnO/ZnO_{1-x}S_x$ core-shell heterostructures, sulfur infiltration approach was used to surface convert ZnO nanorods. For the growth of $ZnO_{1-x}S_x$ over ZnO nanorods was realized by chemiplating process by immersing the ZnO film with nanorod arrays in a 0.1-0.2 M thiourea solution for 4 to16 h duration at 95°C. In this process, sulfide S^{2-} ions generated by slow dissociation of Tu on reaction with Zn^{2+} on ZnO surface partially converts ZnO nanorod into $ZnO_{1-x}S_x$. The composition of sulfur, $0 \leq x \leq 0.2$ is controlled by the duration of the reaction. The substrates were later rinsed in de-ionized water, dried in air and in some cases annealed in N_2 ambient for 1-3 h duration at 100°C.

X-ray diffraction (XRD) patterns of the $ZnO/ZnO_{1-x}S_x$ nanostructure films were recorded on a PANalytical's X'Pert PRO Materials Research Diffractometer with Cu Kα radiation using a Ni filter. Morphology of the films was studied by scanning electron microscopy (SEM) on field emission SEM, Supra 55 VP from Zeiss. The UV-Vis measurements were carried out using Angstrom spectrophotometer in the 250-850 nm range. X-ray photoelectron spectroscopy (XPS) spectra are recorded on PHI 5000 Versaprobe from Physical Electronics.

DISCUSSION

Microstructure of as-formed and chemiplated ZnO nanorods

The microstructure studies show the formation of ZnO nanorod arrays in the as-synthesized ZnO films over quartz substrates. A typical morphology in figure 1(a) shows the ZnO nanorod arrays are of average 200 nm diameter and nearly vertically aligned with lengths 3-5 μm depending on the deposition time. As shown in the high resolution micrograph in figure 1(b), these rods have well defined hexagonal facets along the length with hexagonal shaped top.

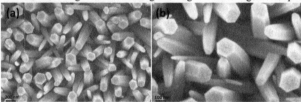

Figure 1. SEM images of ZnO film showing (a) vertically grown nanorods arrays; (b) magnified view showing the hexagonal growth facets of the ZnO nanorods.

EDAX plot in figure 2(a) shows a nearly stoichiometric composition of the nanorods with a Zn:O ratio of 52.8:47.2 at%. Modified microstructure after surface conversion of the ZnO nanorods in the chemiplating process is shown in figure 3. It shows agglomeration of nanocrystals of 10-20 nm average size over the ZnO nanorods. The EDAX plot of the chemiplated ZnO rods shown in figure 2(b) indicates incorporation of sulfur.

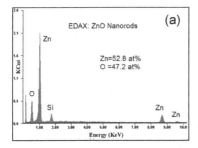 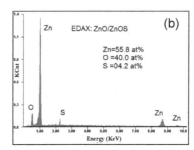

Figure 2. EDAX plots (a) ZnO nanorod arrays; (b) after surface conversion by chemiplating.

Inferred from EDAX analysis, nanocrystals are essentially composed of Zn-O-S compound with the composition after chemiplating for 4 h as Zn-55.8%, O-40.0% and S-4.2%. Typically, the nanocrystals nucleate over the ZnO surface randomly and with progressive chemiplating time these agglomerate and cover the entire surface of the ZnO nanorods creating a coarse Zn-S shell like heterostructure with ZnO nanorods intact in the core. Since the chemiplating solution does not contain Zn source, the structural modification of ZnO nanorods forming a shell of Zn-S compound involves Zn from the ZnO nanorods. It appears that selective diffusion of S^{2-} ions and reaction with Zn^{2+} produces Zn-O-S compound. The EDAX results supported by the XPS studies discussed later show that the nanocrystals are basically zinc oxysulfide ($ZnO_{1-x}S_x$). The $ZnO_{1-x}S_x$ nanocrystals form by self-assembly to dissipate stress cause by diffusion of large ionic radii S in the ZnO lattice of the initial ZnO nanorods. A core-shell $ZnO_{1-x}S_x$/ZnO heterostructure is visualized to form over ZnO nanorods when with progressive growth these nanocrystallites coalesce and form a homogeneous thin sheath over ZnO rods.

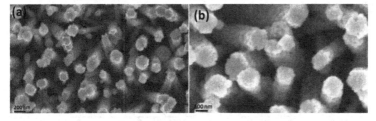

Figure 3. SEM images of ZnO film after chemiplating showing (a) nanocrystal of $ZnO_{0.9}S_{0.1}$ nucleated over ZnO rods; (b) magnified view showing $ZnO_{0.9}S_{0.1}$ nanocrystals of average 20-30 nm size fully covering ZnO nanorods forming a ZnO /$ZnO_{1-x}S_x$ core-shell.

Crystalline Structure of ZnO$_{1-x}$S$_x$/ZnO Core-Shell

The crystal structure of the ZnO nanorods and the ZnOS/ZnO core-shell hetero-structures was determined by x-ray diffraction analysis. A typical diffraction pattern in figure 4(a) confirms a hexagonal crystalline structure of the ZnO nanorods. Highly intense and sharp (002) diffraction line indicates ZnO nanorod arrays grow preferentially with the c-axis oriented normal to the substrate plane and have a narrow size distribution. The XRD pattern of the chemiplated ZnO in figure 4(b) show additional diffraction lines indexed as (100) (002) (101) (110) and (112) planes belonging to the hexagonal wurtzite ZnS phase. Observed shifts in the 2-theta location of (100) (002) (101) compared to the ZnS are consistent with the formation of ZnO$_{1-x}$S$_x$ (x=0.1) as also confirmed by XPS analysis discussed in the next section.

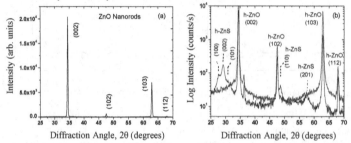

Figure 4. (a) XRD plot of the ZnO film showing hexagonal wurtzite crystal structure of ZnO nanorod arrays with strong c-axis orientation; (b) XRD plot of the ZnO film before and after chemiplating showing formation of Zn-S over ZnO nanorods.

XPS analysis of the ZnO$_{1-x}$S$_x$/ZnO Core-Shell nanostructure

XPS measurements reveal the composition of the chemiplated ZnO nanorods. Surface scan spectra in figure 5(a) show characteristic C1s, Zn2p, O1s and S2p lines. The S2p signal indicates the presence of sulfur on the surface of the ZnO nanorods due to sulfurization reaction with Zn in the chemiplating process. High resolution spectra in figure 5(b) show S2p$_{3/2}$ and S2p$_{1/2}$ peaks positions at binding energy 160.5 and 161.7 eV, which is interpreted as belonging to the Zn-S

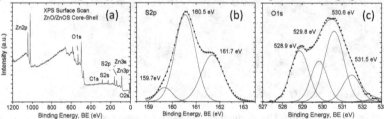

Figure 5. (a) Fast surface scan spectra of ZnO/ZnO$_{1-x}$S$_x$ core-shell nanorods with C1s, S2p, Zn2p and O1s XPS line marked. Unmarked lines are Auger peaks; (b) highly resolved S2p XPS line analysis; (c) highly resolved O1s XPS line analysis.

bonds. The observed XPS peaks are located at a lower binding energy side compared to ZnS values at 161.2 and 162.8 eV [12]. This shift is attributed to a reduced charge on the S-atoms suggesting that S-atoms bond to Zn also share the lattice with more electronegative O-atoms. Figure 4(c) shows high resolution O1s XPS spectra. The O1s XPS line comprises of four separate peaks located at binding energy 528.9, 529.8, 530.6 and 531.5 eV. The most intense line at 530.6 eV is ascribed to oxygen atoms bonded to Zn in wurtzite ZnO single crystal structure. This shows ZnO nanorods are modified only on the surface in the chemiplating process. The two other XPS line resolved at 529.8 and 528.9 eV of lesser intensity are attributed to the oxygen atoms in the lattice position shared with the sulfur atoms indicating the formation of $ZnO_{1-x}S_x$ compound over ZnO nanorods. The XPS data thus suggests the formation of $ZnO_{1-x}S_x$ surface layer over ZnO nanorods forming a core-shell heterostructure by the chemiplating process.

Optical bandgap analysis of the $ZnO_{1-x}S_x$/ZnO Core-Shell nanostructure

Optical band gap of the ZnOS/ZnO nanorods in the core-shell heterostructure was analyzed using the optical transmission spectra. The observed steep change in the optical transmission for the ZnO nanorods in figure 6(a) corresponds to the direct band gap absorption. Optical band gap energy, E_G of the both the samples was determined by fitting the absorption spectra for direct band semiconductors. Figure 6 (b) and (c) shows the experimental plots from which the optical band gap energy was calculated by intercept on the photon energy x-axis. For the ZnO nanorods E_{G1}=3.24 eV is in agreement with ZnO crystal value. The lower band gap E_{G2}=2.78 eV originates from the nanocrystals which form a shell over ZnO nanorods. As discussed earlier, the nanocrystals are basically the ZnO-ZnS alloy. In this case composition of nanocrystals is $ZnO_{1-x}S_x$ (x= 0.2) as determined from the EDAX analysis. It is known that $ZnO_{1-x}S_x$ shows lower band gap energy due to the valance band bowing effect [13]. Valence band bowing is due to the repulsive effect of the S_{3p} orbitals on O_{2p} orbitals which form top of the valence band in ZnO. The optical absorption studies thus confirm the inferences drawn from the microstructure and the XRD studies on the formation of the $ZnO/ZnO_{1-x}S_x$ nanorod core-shell heterostructure.

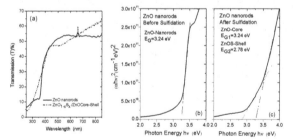

Figure 6. (a) Optical transmission spectra of ZnO nanorods and $ZnO/ZnO_{1-x}S_x$ core shell nanorods. Tauc's plot (b) of ZnO nanorods and (c) of $ZnO/ZnO_{1-x}S_x$ core shell.

CONCLUSIONS

Zinc oxysulfide nanocrystals are grown over ZnO nanorod arrays deposited on seeded quartz substrates by hydrothermal synthesis. With progressive growth of $ZnO_{1-x}S_x$ nanocrystals, full length coverage of the ZnO nanorod leads to the formation $ZnO/ZnO_{1-x}S_x$ core –shell nanostructure with $0 \leq x \leq 0.2$ by reaction time. The band gap engineered $ZnO_{1-x}S_x$ shell and ZnO core nanostructure offers unique advantage in application to solar cells in replacing Cd in non toxic earth abundant materials based thin film solar cells with increased visible wavelength range absorption and low valence band offset and can be used as window layer in solar cells.

ACKNOWLEDGMENT

This project was supported by the Office of Naval Research (ONR) under contract N00014-11-1-0658 which is gratefully acknowledged.

REFERENCES

1. T. Y. Zhai , L. Li , Y. Ma , M. Liao , Xi Wang , X. S. Fang , J. Yao , Y. Bando and D. Golberg, *Chem. Soc. Rev.* **40,** 2986 (2011).
2. J. Schrier, D.O. Demchenko, and L.W. Wang, *Nano Lett.* **7,** 2377 (2007).
3. K. Wang, J. J. Chen, Z. M. Zeng, J. Tarr, W. L. Zhou, Y. Zhang, Y. F. Yan, C. S. Jiang, J. Pern and A. Mascarenhas, *Appl. Phys. Lett.* **96,** 123105 (2010).
4. C. K. Xu, P. Shin, L.L. Cao and D. Gao, *J. Phys. Chem. C* **114,** 125 (2010).
5. S. Y. Bae, H.W. Seo and J. Park, *J. Phys. Chem. B* **108,** 5206 (2004).
6. L.Vayssieres, C.Chaneac, E.Tronc and J.Jolivet, *J.Colloid.Interface Sci.* **205,** 205 (1998).
7. M. Law, L. E. Greene, J. C. Johnson, R. Saykally and P. Yang, *Nature Mater.* **4,** 455 (2005).
8. L. E. Greene, M. Law, D. H. Tan, M. Montano and J. Goldberger, *Nano Lett.* **5,** 1231 (2005).
9. B. Baxter, A. M. Walker, K. V. Ommering and E. S. Aydil, *Nanotechnology* **17,** S304 (2006).
10. B. K. Meyer, A. Polity, B. Farangis, Y. He, D. Hasselkamp, T. Kramer, and C. Wang, *Appl. Phys. Letts.* **85,** 4929 (2004).
11. X. Zhang, Z. Yan, J. Zhao, Zi.Qin and Y. Zhang, *Mat. Letts.* **63,** 444 (2009).
12. D.P. Kim, .I. Kim and K.H. Kwon, *Thin Solid Films* **459,** 131 (2004).
13. Y.Z Yoo., Z. W. Jin, T. Chikyow, T.Fukumura, M. Kawasaki and H. Koinuma, *Appl. Phys. Lett.* **81,** 3798 (2002).

Nanostructures and Nanocomposite Films

Mater. Res. Soc. Symp. Proc. Vol. 1449 © 2012 Materials Research Society
DOI: 10.1557/opl.2012.920

Morphological Studies of Bismuth Nanostructures Prepared by Hydrothermal Microwave Heating.

Oxana V. Kharissova, Mario Osorio, Boris I. Kharisov, Edgar de Casas Ortiz
Universidad Autónoma de Nuevo León, Monterrey, México.
E-mail bkhariss@hotmail.com

ABSTRACT

Elemental bismuth nanoparticles and nanotubes were obtained via microwave hydrothermal synthesis starting from bismuth oxide (Bi_2O_3) in the range of temperatures 200-220°C for 10-45 min. The formed nanostructures were studied by scanning electron microscopy (SEM) and transmission electron microscopy (TEM). Relationship between reaction parameters and shape of the formed nanostructures is discussed.

KEYWORDS: bismuth, nanotubes, truncated nanospheres, microwave hydrothermal synthesis.

1   INTRODUCTION

Metallic bismuth is an important element, having a lot of distinct industrial applications as a component of low-melting alloys, catalysts, for production of polonium in nuclear reactors and tetrafluorhydrazine, among others (Norman, 1997; Cotton et al, 1999). High-purity metal is used, in particular, for measuring super-strong magnetic fields. Bismuth in nanostructurized forms has been mentioned in some monographs (Sergeev, 2006; Koch et al, 2007; Fryxell and Cao, 2007; Asthana and Kumar, 2005; Kharisov et al, 2012), reviews (Soderberg, 2003; Cao and Liu, 2008; Dresselhaus et al, 2003), patents (Penner et al, 2007; Guo, 2006), and a lot of experimental articles. Bismuth nanoparticles, nanopowders, nanowires, nanofilms and other nanostructurized forms have been produced by a host of methods, among which, microwave heating (MW) has been also used to obtain bismuth nanostructures (Kharissova and Kharisov, 2008). Thus, microwave treatment of bulk bismuth in air in a domestic MW-oven (power 800 W and frequency 2.45 GHz) led to formation of bismuth nanoparticles (60-70 nm) with Bi_2O_3 impurities (Kharissova and Rangel, 2007); similar treatment in vacuum (Kharissova et al, 2007; Kharissova et al, 2010) resulted Bi nanotubes formed for 5-15 min. The optimal MW-heating process time was 60 min; the process was found to be highly reproducible and easy. The objective of this report is the study and comparison of various bismuth nanostructures, observed as a result of the synthesis in the conditions of microwave hydrothermal procedure.

2   EXPERIMENTAL

The reactions were carried out in a Teflon autoclave (equipment MARS-5), using bismuth(III) oxide as a precursor and ethyleneglycol (EG) as a reductant. Due to the necessity to exceed boiling point of EG (187°C) and security limits of the equipment, the syntheses were made in the range 200-220°C, reaching pressure close to 300 psi. The reaction times were 10, 15, 30 and 45 min. The formed nanostructures were studied by scanning electronic microscopy (SEM, equipment Hitachi S-5500) and transmission electronic microscopy (TEM, equipment

JEOL 2010-F). The samples were prepared by ultrasonic dispersing of formed products and further application of suspension onto a lacey grid (Lacey Formvar/Carbon, 300 mesh, Copper approx. grid hole size: 63 μm), purchased in *Ted Pella, Inc.* For the observed nanotubes, their diameter (0.7-1.0 nm) was determined applying a *Gatan* analysis (http://www.gatan.com/analysis/).

3   RESULTS AND DISCUSSION

The samples, heated for 10-15 min at 200°C, were analyzed by high-resolution TEM, where 5 nm nanoparticles were observed (Fig. 1a). Fig. 1b shows a nanostructure having 5 nm diameter and length of 58 nm, constituted of various aligned more thin structures.

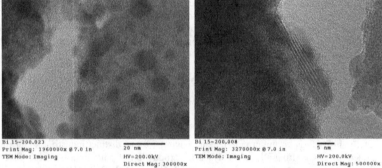

Bi 15-200.023
Print Mag: 1960000x @ 7.0 in
TEM Mode: Imaging
20 nm
HV=200.0kV
Direct Mag: 300000x

Bi 15-200.008
Print Mag: 3270000x @ 7.0 in
TEM Mode: Imaging
5 nm
HV=200.0kV
Direct Mag: 500000x

Fig. 1. TEM images of the formed nanostructures (15 min heating at 200°C). The image shows a) nanoparticles with <10 nm size; the image b) the growth of nanotube agglomerates.

Fig. 2 (in this case, the heating time was 10 min and temperature 220°C) shows high-resolution TEM images of nanoparticles with 15-20 nm diameters, one of which is shown in a larger scale in Fig. 3. The nanoparticles are of a spherical or truncated-spherical form. No nanotubes were observed for 10 min heating.

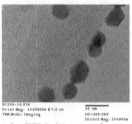

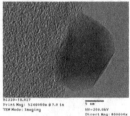

Bi220-10.036
Print Mag: 1640000x @ 7.0 in
TEM Mode: Imaging
20 nm
HV=200.0kV
Direct Mag: 250000x

Bi220-10.027
Print Mag: 5240000x @ 7.0 in
TEM Mode: Imaging
5 nm
HV=200.0kV
Direct Mag: 600000x

Fig. 2. TEM images of samples, heated for 10 min at 220°C.

Fig. 3. High-resolution TEM image of the sample, heated for 10 min at 220°C: nanoparticle with a visible diameter of 18.8 nm.

Further increase of heating time to 30-45 min leads to formation of two types of nanostructures, depending on temperature: spherical nanoparticles are observed in the samples heated at 200°C and 220°C (Fig. 4), as well as multi-wall nanotubes (Fig. 5), observed heating at 220°C only. The maximum observed diameter of spherical nanoparticles reaches 500 nm. For all studied samples with heating time from 10 to 45 min, bismuth atoms were only observed, without oxygen impurity; so, a complete reduction of bismuth oxide to elemental metal nanostructure takes place (Fig. 6).

 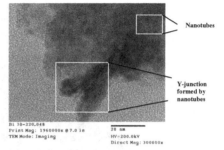

Fig. 4. SEM image of spherical nanoparticles (the sample, heated for 45 min at 220°C).

Fig. 5. TEM image of the sample, heated for 30 min at 220°C.

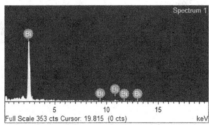

Fig. 6. Elemental analysis of the sample, heated for 10 min at 220°C.

The most interesting results, in our point of view, were observed in case of 15 min heating at 220°C: the formation of blocks, consisting of 12 nanotubes with 0.78-1.08 nm diameter each one (Fig. 7). These blocks have the length starting from 10-15 nm up to 200 nm in several samples and are generally covered by spherical nanoparticles. The nanotubes are perfectly straight and may be confused with nanolines, typical for bismuth (Owen and Miki, 2006). However, we strongly attribute them to individual single-wall nanotubes, connected each other by van der Waals forces, paying attention to visible connections between neighboring "lines" (a thin clearer space is seen between and along tube walls, especially at the end of them, where structural defects appear). Similar results were observed at increase of heating time to 30 min both at 200°C and 220°C. In the last case, the formation of almost Y-junction nanotube conglomerates with diameter of 1.1-1.5 nm each one is seen (Fig. 5).

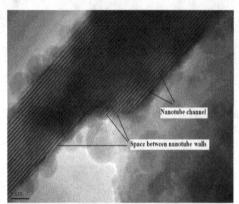

Fig. 7. HRTEM images of the samples, heated for 15 min at 220°C.

Table 1. Summary of the obtained experimental results.

Conditions (reaction time and temperature, °C)	Formed metallic bismuth nanostructures
15 min, 200°C	Nanoparticles with diameter 5 nm and nanotube/nanowire agglomerates with length of 50-60 nm.
10 min, 220°C	Faceted nanoparticles with 15-20 nm diameters. No nanotubes were observed.
15 min, 220°C	Blocks, consisting of 12 nanotubes with 0.6-0.7 nm diameter each one and length starting from 10-15 nm.
30 min, 200°C	Blocks of nanotubes with larger length in comparison with 15 min heating.
30 min, 220°C	Y-junctions of nanotubes with lengths 45-50 nm.
45 min, 200°C	Spherical nanoparticles (diameter up to 500 nm).
45 min, 220°C	Spherical nanoparticles (diameter up to 500 nm) and multi-wall nanotubes.

Comparing the obtained results (a summary is presented in Table 1) with recently reported related investigations, we note that this metal exhibits unusual capacity to form a wide range of nanostructures, in particular nanotubes with different sizes, which are morphologically identical analogues of carbon nanotubes (Rasche and Seifert, 2010; Boldt et al, 2010). In particular, it was concluded in the report (Rasche and Seifert, 2010), dedicated to the stability and electronic properties of Bi nanotubes, that hexagonal prismatic nanotubes of zigzag chirality with radius of 11-30 Å were found to be the most stable nanostructures. In our present investigation, the microwave-formed Bi nanotubes show lesser radius. In any case, in future investigations it would be very useful to establish a dependence of diameter and length of bismuth nanotubes on synthesis method, as well as to try to obtain nanotubes of other elements, which appear in layered allotropes (P, As, and Sb) (Rasche and Seifert, 2010, and cited references therein).[14] At last, we can affirm that the microwave hydrothermal method is an adequate technique for fabrication of bismuth nanostructures.

4 CONCLUSIONS

Elemental bismuth was obtained in the form of nanoparticles and nanotubes in the conditions of microwave hydrothermal heating. Complete reduction of bismuth oxide to metallic bismuth was observed starting from 10 min of treatment. Agglomerates and blocks of metallic nanotubes were observed at intermediate heating times (15-30 min), meanwhile short (10 min) and large (45 min) treatment durations led to spherical and truncated spherical nanoparticles. Additionally, multi-wall nanotubes were observed at large heating times and higher temperature.

5 ACKNOWLEDGMENT

The authors are very grateful to Professor *Miguel José-Yacamán* (University of Texas at San Antonio, TX, USA) and Dr. for access to electronic microscopy equipments at the UTSA Department of Physics & Astronomy.

6 REFERENCES

Asthana, R.; Kumar, A.; Dahotre, N. B. (2005). Materials Processing and Manufacturing Science. 1 edition, Butterworth-Heinemann, 656 pp.

Boldt, R.; Kaiser, M.; Kohler, D.; Krumeich, F.; Ruck, M. (2010). High-yield synthesis and structure of double-walled bismuth nanotubes. Nano Lett., 10, 208-210.

Cao, G. and Liu, D. (2008). Template-based synthesis of nanorod, nanowire, and nanotube arrays. Advances in Colloid and Interface Science, 136 (1), 45-64.

Cotton, F. A.; Wilkinson, G.; Murillo, C. A.; Bochmann, M. (1999). Advanced Inorganic Chemistry, 6th edition, Wiley-Interscience, 1376 pp.

Dresselhaus, M. S.; Lin, Y. M.; Rabin, O.; Jorio, A.; Souza Filho, A. G.; Pimenta, M. A. et al. (2003). Nanowires and nanotubes. Mat. Sci. Engin., C, 23 (1), 129-140.

Fryxell, G. E. and Cao, G. (2007). Environmental Applications of Nanomaterials: Synthesis, Sorbents and Sensors. Imperial College Press, 520 pp.

Guo, T. Nanoparticle radiosensitizers. (2006). Patent WO2006037081.

Kharisov, B. I., Kharissova, O. V., Ortiz-Mendez, U. (2012). Handbook of Less-Common Nanostructures. CRC Press, 863 pp.

Kharissova, O. V. and Rangel Cardenas, J. (2007). The Microwave Heating Technique for Obtaining Bismuth Nanoparticles, in Physics, Chemistry and Application of Nanostructures, World Scientific, pp.443-446.

Kharissova, O.V.; Osorio, M.; Garza, M. (2007). Synthesis of bismuth by microwave irradiation. MRS Fall Meeting. Boston, MA. (November 26-30, 2007). Abstract II5.42. p.773.

Kharissova, O. V. and Kharisov, B. I. (2008). Nanostructurized forms of bismuth. Synth. React. Inorg. Met.-Org. Nano-Met. Chem., 38 (6), 491-502.

Kharissova, O. V.; Osorio, M.; Kharisov, B. I.; José Yacamán, M.; Ortiz Méndez, U. (2010). A comparison of bismuth nanoforms obtained in vacuum and air by microwave heating of bismuth powder. Mater. Chem. Phys., 121, 489-496.

Koch, C.; Ovid'ko, I.; Seal, S.; Veprek, S. (2007). Structural Nanocrystalline Materials: Fundamentals and Applications. 1 edition, Cambridge University Press, 364 pp.

Norman, N. C., Ed. (1997). Chemistry of Arsenic, Antimony and Bismuth. Springer; 1 edition. 496 pp.

Owen, J. H. G.; Miki, K.; Bowler, D.R. (2006). Self-assembled nanowires on semiconductor surfaces. J. Mat. Sci. 41(14), 4568-4603.

Penner, R. M; Zach, M. P., Favier, F. (2007). Methods for fabricating metal nanowires. United States Patent 7220346, http://www.freepatentsonline.com/7220346.html.

Rasche, B.; Seifert, G.; Enyashin, A. (2010). Stability and electronic properties of bismuth nanotubes. J. Phys. Chem. C, 114, 22092-22097.

Sakka, S. (2004). Handbook of Sol-Gel Science and Technology: Processing Characterization and Applications. 1 edition, Springer, 1980 pp.

Sergeev, G. B. (2006). Nanochemistry. Elsevier Science; 1 edition, 262 pp.

Soderberg, B. C. G. (2003). Transition metals in organic synthesis: highlights for the year 2000. Coord. Chem. Rev. 241 (1), 147-247.

Mater. Res. Soc. Symp. Proc. Vol. 1449 © 2012 Materials Research Society
DOI: 10.1557/opl.2012.813

Transparent Film Heaters based on Silver Nanowire Random Networks

Jean-Pierre Simonato[1], Caroline Celle[1], Celine Mayousse[1], Alexandre Carella[1], Henda Basti[1], Alexandre Carpentier[1]
[1]CEA-LITEN / DTNM / LCRE, 17 rue des Martyrs, 38054 Grenoble, France
Email: jean-pierre.simonato@cea.fr

ABSTRACT

We present the fabrication and characterization of transparent thin film heaters (TTFHs) based on silver nanowires. The goal is to develop a simple process for the production of transparent heating elements by large area printing techniques. The TTFHs are based on recently developed random networks of silver nanowires. Thanks to the very low sheet resistance achievable with silver nanowires, we show that it is possible to obtain high heating rates and good steady state temperatures at low voltages, typically below 12 V.

INTRODUCTION

Transparent thin film heaters (TTFHs) represent a considerable market for various applications. Defrosting / defogging of car and plane windows is probably the most well known use of TTFHs. Interestingly, defrosting of windows in airplanes was the first application for TTFHs, permitting high-altitude bombing during World War II.[1] Another typical example is for maintaining LCD displays at optimum operating temperatures for screen response, notably for outdoor LCD displays. Up to now transparent conductive oxides (TCOs) have been widely used for this application, notably zinc oxide derivatives and indium tin oxide (ITO). The properties of TCOs are suitable for most application up to now. However many optoelectronic devices are now evolving to flexible substrates. Many functional systems based on bendable substrates have already been described, including thin film solar cells, OLED or LCD displays, touch screens... For all these devices TCOs can not be considered anymore since they are brittle and thus not resistant to mechanical constraints. Furthermore the manufacturing cost is relatively high, and processes based on indium tin oxide (ITO) are somehow risky because the price of indium is particularly fluctuating due to its scarcity, its increasing use and the still insufficient recycling process.

Several innovative approaches to obtain flexible TTFHs have been developed these last years, in particular with carbon nanotubes [2-7] and grapheme [8, 9]. Nevertheless the sheet resistivity of the networks made with these materials at high transparency is still too high for low voltage applications.

At the same time, new transparent conductors have been realized thanks to metallic nanowires. In particular silver nanowires (Ag NWs) are rather easy to make in large amount from batch solution reactions.[10-14] Performances of Ag NWs for conducting electrons at high transparency are excellent with very low sheet resistances, comparable to those of ITO (< 30 ohm.sq^{-1}).[14-19] We present herein results on new electrothermal film heaters based on random networks of silver nanowires and we demonstrate that excellent heating performances can be obtained at high transparency.

RESULTS AND DISCUSSION

The study was mainly divided into three steps. The nanowires were first synthesized in solution and characterized, then random networks of Ag NWs were prepared on different substrates and finally heating performances were monitored at various applied voltages.

Synthesis of silver nanowires

The silver nanowires were synthesized by the polyol method[13] and purified to remove remaining nanoparticles (see Figure 1). The mean length of nanowires was ~8 μm and diameters in the 50-80 nm range. It is important to obtain high aspect ratio nanowires because it allows to fabricate percolating random networks of nanowires with very good ratio between transparency and electrical conductivity. Before the deposition step, Ag NWs were dispersed either in water or organic solvents like alcohols, for instance isopropanol. The solutions of silver nanowires were stable for weeks.

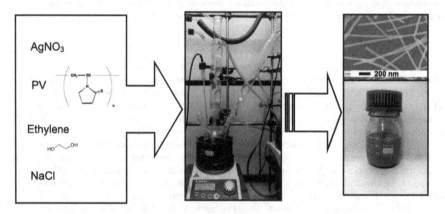

Figure 1. Batch synthesis of nanowires from silver salt

108

Fabrication of random networks of silver nanowires

Various techniques can be used to fabricate transparent electrodes including spin-coating, airbrush spray, drop casting and others. Glass was used for these first experiments but flexible substrates could certainly also be used without any obvious problem. Indeed, It is already known that random networks of nanowires can be transferred by PDMS stamps, laminated or embedded in polymers to afford conductive transparent materials with lower roughness.[17 20, 21] Typically, sheet resistances of few tens ohm.sq^{-1} are obtained at 90% transparency in the visible spectrum, down to less than 1 ohm.sq^{-1} at ~60% transmission. These performances are in the same range of those obtained with ITO.

In this study, TTFHs were obtained by spincoating solutions of Ag NWs onto Eagle XG glass substrates. The sheet resistances of the electrodes were measured by a four probe resistivity meter and the transmittance by a UV-vis-NIR spectrophotometer. The power dissipated in a resistive conductor can be described by the Equation $P=V^2/R$, where V is the applied voltage and R the total resistance, thus it can be understood that with low sheet resistance high heat dissipation can be obtained at fixed bias. And precisely, this is a specific advantage of metallic nanowires networks to afford very low sheet resistances at high transparency.

Figure 2. TTFHs deposited on glass and placed upon the printed CEA logo.

Heating performances

We prepared the TTFHs by applying silver paste to make a low-resistance contact on two opposite sides of the film. We built an experimental setup for monitoring the temperature increase while DC voltage is applied (Figure 3).

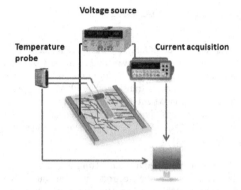

Figure 3. Experimental setup for monitoring heat generation of TTFHs at various voltages.

We applied incremental input voltages ranging from 3 to 12V by a DC power supply (Agilent 34410A). We measured the electrical current through the TTFHs and we monitored the heating effect with a thermocouple. Figure 4 shows the temperature plot as a function of time, for a 88% transparent TTFH having a sheet resistance of 80 ohm.sq^{-1}.

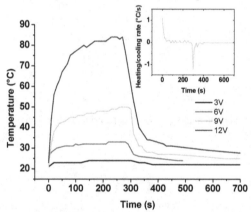

Figure 4. Temperature as a function of time for a 80 ohm.sq^{-1} TFH on glass at different applied voltages. Insert: derivative of the temperature vs time at 9V applied voltage.

Steady-state temperatures were significantly different and correlated with applied biases. This is due to the generation of heat by the passage of electricity through a resistance, *i.e.* Joule effect, which is proportional to the square of the applied voltage: $P=V^2/R$. This fits well with observed steady state temperatures, even by taken into account that resistance is slightly modified by increasing the temperature. As shown in the insert, the heating/cooling rates calculated from the derivative of the temperature with respect to time $(\partial T / \partial t)$ was slightly above 1 °C.s^{-1} at an applied voltage of 9V. This is interesting for many applications requiring high heating rates at low voltage.

These results compare very well with other TTFHs based on carbon nanotubes or graphene. Thanks to the low sheet resistance achievable with random networks of silver nanowires, it is possible to reach high temperatures at low voltage, typically below 12 V, which might be for instance of interest for electrical defrosting in automobile systems.

Concerning the mechanical stability of the TTFHs, we did not observe delaminating between the film and substrate even after 10 heating-cooling cycles. However, the film has poor adhesion onto glass and further experiments are in progress to improve bond between the two materials.

CONCLUSIONS

In summary, we developed a new method to fabricate high performance TTFHs based on random networks of Ag NWs. We have shown that it is possible to fabricate film heaters that may be very efficient even at low voltages. This is mainly ascribed to the excellent electrical properties of transparent conductors made with random networks of Ag NWs. To our knowledge, this is the first report of TTFHs based on metallic nanowires. We think these results could be a useful approach for the engineering of TTFHs, and it opens the way to fabricate highly bendable TTFHs by depositing the networks of Ag NWs onto various transparent flexible substrates such as polyimide or polyester plastic foils.

ACKNOWLEDGMENTS

Authors thank "Carnot Energies du Futur" for fundings.

REFERENCES

[1]   R. G. Gordon, *MRS Bulletin,* pp. 52-57, 2000.
[2]   Y. H. Yoon, J. W. Song, D. Kim, J. Kim, J. K. Park, S. K. Oh, and C. S. Han, *Advanced Materials,* vol. 19, pp. 4284-4287, 2007.
[3]   D. Kim, H.-C. Lee, J. Y. Woo, and C.-S. Han, *The Journal of Physical Chemistry C,* vol. 114, pp. 5817-5821, 2010.
[4]   H.-S. Jang, S. K. Jeon, and S. H. Nahm, *Carbon,* vol. 49, pp. 111-116, 2011.
[5]   Z. P. Wu and J. N. Wang, *Physica E: Low-dimensional Systems and Nanostructures,* vol. 42, pp. 77-81, 2009.
[6]   T. J. Kang, T. Kim, S. M. Seo, Y. J. Park, and Y. H. Kim, *Carbon,* vol. 49, pp. 1087-1093, 2011.
[7]   E. J. Spadafora, K. Saint-Aubin, C. Celle, R. Demadrille, B. Grévin, J.-P. Simonato, *Carbon,* 10.1016/j.carbon.2012.03.010, 2012.

[8]   J. Kang, H. Kim, K. S. Kim, S.-K. Lee, S. Bae, J.-H. Ahn, Y.-J. Kim, J.-B. Choi, and B. H. Hong, *Nano Letters,* vol. 11, pp. 5154-5158, 2011.

[9]   D. Sui, Y. Huang, L. Huang, J. Liang, Y. Ma, and Y. Chen, *Small,* vol. 7, pp. 3186-3192, 2011.

[10]   Y. Sun, B. Gates, B. Mayers, and Y. Xia, *Nano Letters,* vol. 2, pp. 165-168, 2002.

[11]   L. Hu, H. S. Kim, J.-Y. Lee, P. Peumans, and Y. Cui, *ACS Nano,* vol. 4, pp. 2955-2963, 2010.

[12]   A. Madaria, A. Kumar, F. Ishikawa, and C. Zhou, *Nano Research,* vol. 3, pp. 564-573, 2010.

[13]   S. Coskun, B. Aksoy, and H. E. Unalan, *Crystal Growth & Design,* vol. 11, pp. 4963-4969, 2011.

[14]   D.-S. Leem, A. Edwards, M. Faist, J. Nelson, D. D. C. Bradley, and J. C. de Mello, 4*Advanced Materials,* vol. 23, pp. 4371-4375, 2011.

[15]   X.-Y. Zeng, Q.-K. Zhang, R.-M. Yu, and C.-Z. Lu, *Advanced Materials,* vol. 22, pp. 54484-4488, 2010.

[16]   L. B. Hu, H. Wu, and Y. Cui, *MRS Bulletin,* vol. 36, pp. 760-765, Oct 2011.

[17]   A. R. Madaria, A. Kumar, and C. W. Zhou, *Nanotechnology,* vol. 22, p. 7, May 2011.

[18]   V. Scardaci, R. Coull, P. E. Lyons, D. Rickard, and J. N. Coleman, *Small,* vol. 7, pp. 2621-2628, 2011.

[19]   Z. Yu, Q. Zhang, L. Li, Q. Chen, X. Niu, J. Liu, and Q. Pei, *Advanced Materials,* vol. 23, pp. 664-668, 2011.

[20]   Z. Yu, L. Li, Q. Zhang, W. Hu, and Q. Pei, *Advanced Materials,* vol. 23, pp. 4453-4457, 2011.

[21]   W. Gaynor, G. F. Burkhard, M. D. McGehee, and P. Peumans, *Advanced Materials,* vol. 23, pp. 2905-2910, 2011.

Mater. Res. Soc. Symp. Proc. Vol. 1449 © 2012 Materials Research Society
DOI: 10.1557/opl.2012.1173

Synthesis, Characterization and Water Vapor Sensitivities of Nanocrystalline SnO2 Thin Films

M. Chacón, A. Watson, I. Abrego, E. Ching-Prado*
Natural Science Department, Faculty of Science and Technology
Technological University of Panama, Panama

J. Ardinsson
Development Center of Nuclear Technology (CDTN),
Federal University of Minas Gerais, Bello Horizonte, Brasil

C. A. Samudio Pérez
University of Passo Fundo, Porto Alegre, Brasil

Abstract

Thin films of SnO2 were prepared via wet chemical method and deposited by dip-coating on glass substrate. The annealing temperatures of the samples were 300, 400, 500 and 600° C, respectively. Water vapor sensor responses were measured and the experimental results are tested using the Freundlich model. The better water vapor sensitivities were obtained for annealing temperatures of 500 and 600°C, respectively. The samples were characterized morphological and structurally by SEM, XRD, Mössbauer spectroscopy and Raman spectroscopy. The fringes features in the ultraviolet-visible region indicate films thickness around 370 nm. The results are discussed in terms of the fine grain size of the samples.

Introduction

Research in semiconductor metal oxide nanocrystalline thin films is of great importance due to its wide range in technological applications, such as optoelectronic devices, fabricating solar cells, electrode materials for Li-batteries, and solid state gas sensor, among many others. Tin oxide (SnO2) is one of these materials that have been utilized as gas sensing due to a combination of its physical and chemical properties [1].

A disadvantage of currently conventional SnO2 gas sensors is that they typically operate at temperatures higher than 300°C, which may cause structural changes and sensor instability. So that, fabricating sensing materials with low operating temperatures is desirable, because it permit the sensor work properly and reduce the energy consumption [1]. Furthermore, since the sensing mechanism of gas sensors is based on the chemisorption reactions that take place at the surface of SnO2, so increasing specific surface area of the sensitive materials leads to more sites for adsorption of surrounding gases. Some recent studies have been concentrated on improving gas sensitivity as well as reducing operating temperature by decreasing SnO2 size to nanoscale [1,2]. On the other hand, some important advantages of SnO2 are its stability in air, relative cheapness, and simplicity of preparation. Thus, tin oxide thin films have been grown for many relative no expensive preparation techniques, such as: sol gel, evaporation, hydrothermal, spray pyrolysis, etc [3]. Several behaviors of these tin oxide thin films have been observed to change by modifying the preparation method and environment condition. Actually, the use of polycrystalline SnO2 as solid state gas sensor is of great interest with regard to the relationship between electrical properties and crystalline size. In this paper nanocrystalline SnO_2 thin films grown by liquid chemistry with samples annealed to 300, 400, 500, and 600° C were investigated. Electrical responses to presence of water vapor concentration were registered at room temperature (RT). Also, the samples were studied by SEM, XRD, Mössbauer spectroscopy and Raman spectroscopy.

Experimental

Pure SnO_2 solution were prepared starting from commercially available anhydrous tin (IV) chloride $SnCl_4$, distilled water, propanol (C_3H_7OH), and isopropanol ($2-C_3H_7OH$) in the following molar ratio:$1(SnCl_4) : 9(H_2O) : 6(C_3H_7OH) : 6(2-C_3H_7OH)$. In order to growth thin films, the main solution was diluted to 60 % in propanol [3]. The films were deposited by dip-coating on glass. The as-deposited thin film samples were dried at 80° C for one hour in the open air using a hotplate. Through a resistive furnace, the samples were annealed in air for 2 hours at 300, 400, 500 and 600°C, respectively.

Silver paint electrodes were used at a separation of about 5.0 mm for the study of electrical response to water vapor concentration of all samples. The micro-gas sensors were placed inside one gallon measurement chamber and different volume of distilled water were injected, using a micropipette, directly to a filter flask, which it is on a hot plate at constant temperature higher than 100°C. Dry air with a flow rate of 6 liters per minute was used as a carrier water vapor.

The surface morphologies were analyzed at room temperature using a Scanning Electron Microscope (SEM) ZEISS EVO 40 VP, while the structural properties were recorded, at room temperature, by an X-ray diffractometer Siemens D-5000, a Raman spectrometer Delta Nu Advantage 532, and a Mössbauer spectrometer. Using a Spectronic Genesys 5, the fringes features at ultraviolet-visible region were utilized for calculate the thickness of the samples.

Theory

To describe the interaction between surface resistance based in MOS sensor, like SnO_2, and adsorbed gas several isotherm models have been employed to fit experimental data, such as Langnuir, Scatchard, Brunauer-Emmett-Teller, Freundlich, Langmuir–Freundlich, and others. The Freundlich is an empirical expression that takes into account the heterogeneity of the surface and it can be expressed as follows [4]:

$$\frac{R^{air}}{R^{gas}} = 1 + zn^m \qquad (1)$$

Where $z = \alpha\gamma^m$ and $\gamma = \dfrac{n^{eq}}{n}$.

R^{gas} corresponds to the electrical resistance in presence of the gas, R^{air} is the electrical resistance in air (baseline), n is the initial gas mole number, n^{eq} is the gas mole number at thermodynamic equilibrium, α is a parameter which includes electrical and thermodynamic quantities, and m is a coefficient.

In many metal oxide materials, the electrical resistance R^{air} is very large (depending electrodes distance, etc.) in comparison with the maximum instrumental limit, so that a possible solution is connect a resistance R_p in parallel to the oxide sensor in order to register electrical response. Thus, if P variable is defined as $P = R_{air}^{mea} - R_{gas}^{mea}$, where R_{gas}^{mea} and R_{air}^{mea} are the measured resistances and they have the usual meaning, then:

$$P = R_{air}^{mea} - \frac{R_p}{\left(\dfrac{R_p}{R_{air}^{mea}} - 1\right)\left(1 + zn^m\right) + 1} \qquad (2)$$

It can be noted that P is proportional to the sensor sensitivity, being the sensitivity as that usually defined in the literature. In addition, it can be observed that the Freundlich model does not consider adsorption saturation.

Results and Discussion

Figure 1 shows the SEM micrograph of the SnO_2 thin films prepared at 300, 400, 500 and 600°C. SEM pictures indicate that the films are very dense, with clear evidences of cracking for the samples annealed at 300 and 400°C. Also, the picture presents a grain size distribution in all the samples. A significant increase in grain size can be observed when the annealed temperature changes from 300 to 400°C. However, a little increase in grain size dimension is found for the films prepared at 500 and 600°C in comparison to those in the sample grown at 400°C, where the picture suggests a very fine grain, around 50 nm or less.

The ultraviolet-visible fringes features indicate a film thickness value around 370 nm.

Figure 2 shows a typical X-ray diffractogram found in all the samples, where it is possible observe the (110), (101) and

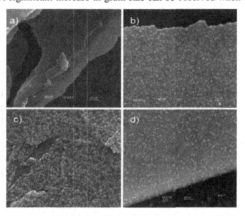

Figure 1: SEM micrographs of SnO_2 films annealed at a) 300, b) 400, c) 500 y d) 600°C

(211) crystal planes [3]. As expected, the X-ray results indicate that the crystallization degree increase with increasing annealing temperature. It confirms the tetragonal structure of the samples associated to SnO_2 formation [5]. Also, this is supported by Mössbauer spectroscopy measurements realized to all the films; where Figure 3 shows a isomer shift around 0,01 mm/s and a quadrupole splitting close to 0,51 mm/s [6]. They are characteristic of polycrystalline SnO_2 tetragonal structure. In addition, a Raman spectrum of thicker film (not presented here), prepared in similar way than the 600°C thin film, shows the A_{1g} and B_{2g} phonon bands characteristic of SnO_2 powder material.

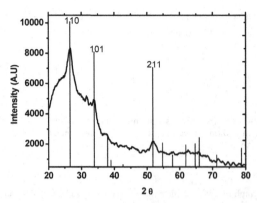

Figure 2: X-ray diffractogram of SnO_2 thin film prepared at 600°C.

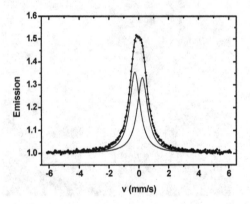

Figure 3: A typical Mössbauer spectrum obtained in the thin film samples.

The figure 4 presents room temperature (RT) sensor response as function of water vapor concentration for the thin film prepared at 600°C. Two regions could be clearly identified in this experimental band. They correspond to the adsorption region and the desorption region, respectively. The former is associated to the adhesion of water molecules on the thin films surface sites, and the latter is the removal of water molecules on these surface sites, caused by dry air flow. The figure indicates that the electrical response variation P increase with increasing the water vapor content in the chamber. Also, sensor test realized to all the films show that higher water vapor sensitivity is obtained in the samples prepared with higher annealing temperature. Thus, a good electrical response to the presence of water vapor is found for the thin films annealed at 500 and 600°C. The figure 5 presents the electrical response for the thin film prepared to 500°C as water volume content is equal or greater than 200

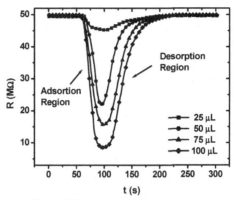

Figure 4: RT water vapor electrical response for the sample prepared at 600°C.

µL. It can be noted that the electrical response variation P almost does not changes as water volume increase from 200 to 400 µL. However, the widths of this electrical band increases significantly as increasing water volume. Thus, a longer flat section in the bottom of the band, called saturation region, seems to take place as water volume increase. Therefore, the sensor test result could be explained of the following form: In each sample exist a permissible number of surface sites that can be occupied by water molecules in the adsorption process.

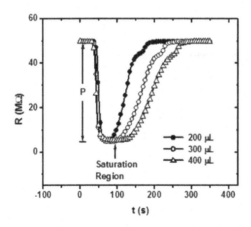

Figure 5: RT electrical response for the sample prepared at 500°C in presence of high water content.

Thus, when water is injected in the chamber, a certain amount of these surface sites are occupied and an electrical response variation P, different than cero, is obtained. If the water volume is increasing more sites are involved and P value increase. However, for some value of water volume, called here critical volume, all sites are occupied and a maximum P value (P_{max}) is measured. So that, if the injected water volume is higher than the critical one, then the same P_{max} is obtained, because it also involves all the sites. On the other hand, the excess water molecules, as compared to those in critical volume, will be in thermodynamic equilibrium over the sample. This causes the desorption process to take place after removal the excess water molecules using a dry air flow. Thus, the time to remove excess water molecules corresponds to the saturation region.

Figure 6 shows that the electrical response area is proportional to water volume. Figure 7 corresponds to the P value fitted using eq. (2). A very good fitting is obtained with $m = 2.0$.

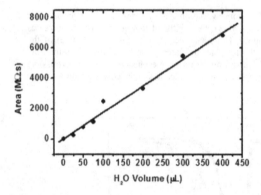

Figure 6: RT electrical response area as function of water volume content.

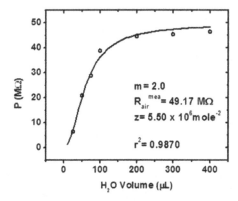

Figure 7: RT electrical response variation P as function of water volume. The solid line corresponds to the fitting using eq.(2).

Conclusion

SnO_2 Thin films were prepared by wet chemical method on glass substrate. Grain sizes of 50 nm or less were found in the samples. The electrical response variation area can be used as a water sensor parameter, while electrical response variation P is a measured of the surface sites number available for the water molecules adsorption process. The P value results, just before of saturation, can be explained very well using a Freundlich isotherm model.

Acknowledgements

This work was partially supported by FID05-061 and APY-GC-10-046A SENACyT grants. Thanks to SmithSonian Tropical Research Institute, especially to Jorge Ceballos, for collaborate in SEM measurements. Also thanks to Nanostructure Group from the Institute of Materials Jean Rouxel-CNRS, France, for help in XRD experiments.

*Corresponding author at: eleicer.ching@utp.ac.pa.

References

1. Qi-Hui Wu, Jing Li and Shi-Gang Sun, Current Nanoscience, **6,** 525-538 (2010).
2. M. Batzill and U. Diebold, Progress in Surface Science, **79**, 47–154 (2005).
3. P. Siciliano, Sensors and Actuators, **B70**, 153–164 (2000).
4. A. Abbas and A. Bouabdellah, Sensors and Actuators, **B145**, 620–627 (2010).
5. M. Bhagwat, P. Shah, and V. Ramaswamy, Materials Letters, **57,** 1604–1611 (2003).
6. P. Bussiere, Review physics application, **15**, 1143-1147 (1980).

Mater. Res. Soc. Symp. Proc. Vol. 1449 © 2012 Materials Research Society
DOI: 10.1557/opl.2012.1041

Synthesis of Crystalline ZnO Nanosheets on Graphene and Other Substrates at Ambient Conditions

Phani Kiran Vabbina[*a], Santanu Das[b], Nezih Pala[a], Wonbong Choi[b]

[a]Electrical & Computer Engineering Department, Florida International University, Miami, FL.

[b]Mechanical and Materials Engineering, Florida International University, Miami, FL.

ABSTRACT

We report on the fabrication of ZnO nanosheets on Graphene and other substrates at ambient conditions. The growth mechanism and the effect of the substrate are also discussed. Our synthesis method is based on sonochemical reaction of Zinc nitrate hexahydrate and hexamethylenetetramine in aqueous solutions. Extensive analysis by transmission electron microscopy, energy dispersive x-ray spectroscopy (EDS) revealed crystalline ZnO composition of the ZnO nanosheets. The proposed method is a rapid, inexpensive, low-temperature, catalyst-free, CMOS compatible and environmentally benign alternative to existing growth techniques.

INTRODUCTION

Graphene has attracted a lot interest due to its unique electrical and optical properties. Growing oxides over graphene will provide additional functionality to graphene in electronic and optoelectronic applications. This will also enable us to integrate Graphene in semiconductors [2]. Zinc Oxide with its wide band gap of 3.37 eV, large exciton binding energy and high thermal stability has attracted wide research recently. ZnO nanostructures are expected to find applications in light emitting and detection devices in UV-visible spectral range, energy harvesting devices and in gas sensors. 2-D ZnO nanosheets with their large surface area and flat topography are among those nanostructures. ZnO nanosheets have been successfully synthesized by various methods such as vapor-solid and hydrothermal methods [9, 10]. However to be able to use in practical low cost electronic applications, these techniques are either not cost effective or the conditions are too harsh to give any flexibility for the growth on desired substrates. In this regard, we have recently reported on a sonochemical method which involves use of high intensity ultrasonic irradiation of an aqueous solution as an alternative method for the synthesis of ZnO nanowires and nanowalls [1]. Our method resulted in a ten times faster synthesis of ZnO nanostructures compared to the more conventional approaches such as hydrothermal synthesis, due to the fast hydrolysis rate caused by imploding cavitation bubbles [1] and does not require a seeding layer like previously reported sonochemical synthesis methods [5].

Here we report the synthesis of ultra-thin ZnO nanosheets on Graphene and other substrates by simple sonochemical method and present extensive crystallographic and surface and electrical studies by TEM, EDS, and AFM. With the advantages of being inexpensive, low-temperature, catalyst-free, CMOS compatible and environmentally benign, our reported method has the potential of being used for many electronic and photonic applications.

EXPERIMENT:

ZnO nanosheets were synthesized in a custom made quartz container with a built-in sample holder. All the reagents used in experiments were purchased from Sigma-Aldrich with analytical grade and no further purification was done. A solution of 0.3M ZNH and 0.2M hexamethylenetetramine ($CH_2)_6N_4$) (HMT) was prepared at room temperature by stirring with a magnetic stir bar at 350 rpm for 5 minutes to ensure a mixed solution was used as the only solution for the growth of ZnO nanosheets. The substrates were cleaned with methanol, acetone isopropanol and DI water before immersing it into the solution. The solution with the immersed substrate then was irradiated using a commercially available high intensity ultrasound setup (750W ultrasonic processor, Sonics and Systems), during 2-6 minute cycles with an amplitude of 70% of the maximum amplitude (~30 W.cm^{-2}). The global temperature of the aqueous solutions did not exceed ~70°C. SEM imaging of the synthesized nanostructures was performed using a JEOL6335. A 200kV Phillips CM200 was used for TEM analysis. Multimode Nanoscope IIID from Bruker (Veeco) was used to study the sheet thickness and surface roughness measurements. Different substrates were used to test the substrate effects on ZnO morphologies. In order to understand the growth mechanism, ZnO sheets were first grown on Si, SiO$_2$/Si, Copper, Aluminum, Quartz, Ni and Fe alloys for varying lengths of times. After successful synthesis of ZnO nanosheets on these substrates and the growth mechanism studied, the process was carried out on Graphene over Copper, SiO$_2$, Si and Poly ethylene tetrapthalate.

RESULTS & DISCUSSION

<u>**ZnO Nanosheets Growth**</u>

ZnO nanosheets were successfully grown on Si, Si/SiO$_2$, Quartz, PET, alloy of Ni and iron. It was observed that on Al and Cu, the growth conditions favored wall growth than sheet growth. To understand the growth kinetics and growth mechanism, growth of ZnO nanosheets on Si substrate at different times is observed as shown in figure 1. As we see the nanosheets grow from irregular shape at 1 min. to hexagonal and square shaped highly transparent and thin nanosheets at 4 and 12 minutes. It can also be observed that these sheets grow in layers with new layers forming over the older ones. The size and density of these sheets increases with time and number of cycles, thus covering the whole substrate as seen from figure 1(c). The composition of the nanosheets formed is verified by EDS. As seen from figure 2 the nanosheets formed are purely ZnO and we see no other impurity except for the Si which comes from the substrate and low percentage Carbon from measurement setup.

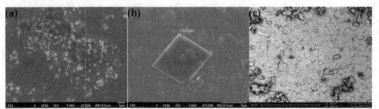

Figure 1. (a) Growth of thin irregular sheets on Si after 1 minute growth cycle **(b)** Growth after 4 minutes of growth cycle **(c)** Growth after 2 cycles of 12 minutes, covering the whole substrate.

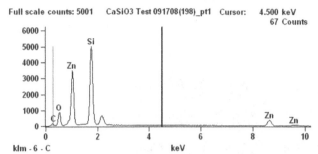

Full scale counts: 5001   CaSiO3 Test 091708(198)_pt1   Cursor: 4.500 keV

Figure 2. The measured EDS analysis shows the purity of the ZnO sheets grown on Si substrate.

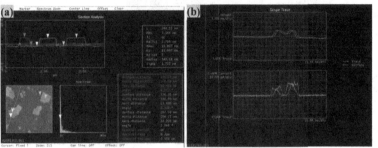

Figure 3. AFM studies of **(a)** thickness, surface roughness and **(b)** conductivity of ZnO nanosheets on Si substrate.

The thickness of the sheets was measured by AFM. It was observed that the sheets are extremely thin with thickness ranging from 5nm to 15 nm as shown in figure 3. The sheets are extremely flat and smooth with mean roughness around 0.23 nm. To further understanding of the crystal structure and growth, structural analysis is performed by using TEM

Figure 4(a) shows a single sheet in TEM. The SAD analysis of the sheets shown in figure 4(b) reveals a single crystal nature with wurtzite structure of the nanosheets. From the SAD pattern it can be seen that the sheets are oriented in (0002) plane. From figure 1 (a)-(c) which show the SEM pictures at different times we observe that the sheets grow from irregular shaped thin sheets to rectangular shaped single crystal sheets in very quick time. We suspect the high

growth rate along $(01\bar{1}0)$ and $(0\bar{1}10)$ planes instead of (0001) plane could be the reason, further crystallographic study is required to confirm this. We have earlier reported growth of rods [1] in which HMT plays a key role by inhibiting the growth in non-planar facets while facilitating the growth in (0002) plane [7, 8]. Earlier when we reported ZnO nanorod growth we used equal concentrations of ZNH and HMT. Here the concentration of HMT in comparison with ZNH is decreased. We use 0.2M HMT and 0.3 M ZNH. This decrease in the concentration of HMT, resulted in the growth of the nanosheets. Generally in wurtzite structure the growth rate along the [0001] direction is higher, however under certain circumstances, the planes having lower specific surface energy are preferred [4, 6, 8]. It is evident from the pictures in figure 1a-c, the growth along [01$\bar{1}$0], [0$\bar{1}$10] directions are preferred. This leads to hexagonal and with enough time predominantly square shaped sheets.

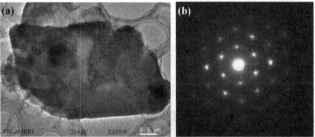

Figure 4. (a) TEM picture of single ZnO nanosheet on Si **(b)** SAD pattern confirming the crystalline structure and showing (0002) orientation

ZnO Nanosheets Growth on Graphene:

Graphene samples were grown by thermal chemical vapor deposition (CVD) of methane under reducing atmosphere at 1000°C. Latter, as grown CVD graphene has been transferred on to SiO$_2$/Si, Si, PET, Cu substrates via a chemical process as reported elsewhere [11]. From figure 5 we observe a consistent time dependent growth of nanosheets on Graphene. Comparing figure 5(a) with 5(b) we can see that the density of growth is higher on Graphene over SiO$_2$ compared to growth on bare SiO$_2$. We also observe that the density increases with time by comparing figure 5(a) with figure 5(c). However, in case of Copper, we observed growth of nano walls as seen in figure 5(d). From this we can conclude that the morphology of the ZnO sheets on Graphene is influenced by the substrate on which Graphene is grown as well. The presence of the Graphene on the samples is confirmed by Raman spectroscopy as shown in figure 6(a). The HRTEM and SAD data shown in figure 6(b)-(d) confirms the single crystalline nature of the ZnO sheets grown on Graphene, with hexagonal wurtzite structure and oriented in (0001) plane.

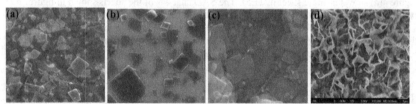

Figure 5. **(a)** ZnO nanosheets grown on Graphene on SiO$_2$/Si. Process time is 3 minutes. **(b)** ZnO nanosheets grown on SiO$_2$/Si, for 3 minutes. **(c)** ZnO nanosheets grown on Graphene over PET for 15 mins. **(d)** Growth on Graphene over Copper for 15 mins, we observe wall growth.

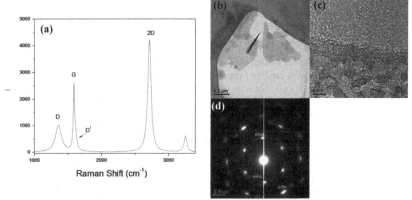

Figure 6. (a) Raman shift showing the presence of Graphene (b) TEM picture of ZnO sheets on Graphene, (c) HRTEM picture of ZnO nanosheets on Graphene (d) SAD pattern for nanosheets on Graphene.

Current- voltage characteristics of ZnO nanosheets are shown in Figure 7(a). The measurements were taken on nanosheets grown on Si and Graphene/Si substrates by a 15 minute growth cycle. The comparison of the I-V curves shows the increase in conductivity of nanosheets when grown on Graphene. An increase in conductivity in presence of UV light is observed which is a characteristic of ZnO.

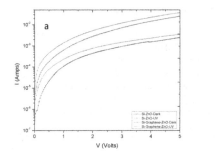
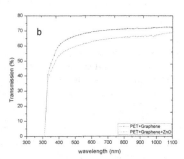

Figure 7. (a) I-V curves for ZnO nanosheets without UV light and with UV light on Si and Graphene/Si substrates (b) Percentage transmission of Graphene on PET and ZnO nanosheets on Graphene over PET

The transmission characteristics were measured by growing ZnO nanosheets on multilayer Graphene over PET. As seen in figure 7(b). The reduction in transmission due to ZnO nanosheets is very small close to 5%. The lower average transmission in this case is due to multilayer Graphene. The reduction in transmission below 300nm confirms the presence of ZnO.

CONCLUSION

We have reported a sonochemical method to develop ZnO nanosheets on Graphene and other substrates. The grown structures are highly pure and single crystal in nature with orientation along c-axis i.e. (0002) plane, compared to the normal process the growth is faster in the $(01\bar{1}0)$ and $(0\bar{1}10)$ planes, resulting in flat hexagonal sheets which grow in to square sheets. The sheets are extremely thin with thickness ranging between 5nm and 15nm. From current voltage characteristics and the transmission values, we can see that the growth of ZnO nanosheets over Graphene resulted in 100 fold increase in conductivity while the transmission values were not affected much. The proposed method is a rapid, inexpensive, low-temperature, catalyst-free, CMOS compatible and environmentally benign alternative to existing growth techniques and integrating Graphene with ZnO nanosheets can find applications in fields like dye-sensitized solar cells and opto-electronic devices.

ACKNOWLEDGEMENT

The work was supported by US National Science Foundation CAREER award program (monitored by Samir El-Ghazaly).

REFERENCES

1. Phani K. Vabbina. Et al., "Synthesis of crystalline ZnO nanostructures on arbitrary substrates at ambient conditions", Proc. SPIE 8106OH (2011)
2. Jian Lin. et al., "Heterogeneous Graphene Nanostructures: ZnO Nanostructures Grown on Large-Area Graphene Layers", Small Vol 6, 2448-2452, 2010
3. J. Q. Hu. Et al., "Two-dimensional micrometer-sized single-crystalline ZnO thin nanosheets", applied physics letters Vol 83 November 2003.
4. Soumtra Kar. et al., "Simple Solvothermal Route To Synthesize ZnO Nanosheets, Nanonails, and Well-Aligned Nanorod Arrays", Journal of Physical Chemistry B 2006, 110
5. Avinash P. Nayak, et al. "Purely sonochemical route for oriented zinc oxide nanowire growth on arbitrary substrate", Proc. SPIE 768312 (2010).
6. kar, S. et al., "One-Dimensional ZnO Nanostructure Arrays: Synthesis and Characterization", Journal of Physical Chemistry B 2006, 110
7. Sheng Xu, Zhong Lin Wang, "One-Dimensional ZnO Nanostructures: Solution Growth and Functional Properties". Nano Research Vol. 4 1013-1098, 2011.
8. Sunandan Baruah, Joydeep Dutta, "Hydrothermal growth of ZnO nanostructures" Science and Technology of Advanced Materials, Vol. 10 2009
9. Run Liu. et al, "Epitaxial Electrodeposition of Zinc Oxide Nanopillars on Single-Crystal Gold" Chemistry of Materials, 2001, 13(2)
10. Jianqiang X. et al, "Hydrothermal synthesis and gas sensing characters of ZnO nanorods" Sensors and Actuators B 113, 526- 531, (2006)
11. Das, S.; Sudhagar, P.; Verma, V.; Song, D.; Ito, E.; Lee, S. Y.; Kang, Y. S.; Choi, W., Amplifying Charge-Transfer Characteristics of Graphene for Triiodide Reduction in Dye-Sensitized Solar Cells. *Advanced Functional Materials* 2011, 21, 3729-3736

Mater. Res. Soc. Symp. Proc. Vol. 1449 © 2012 Materials Research Society
DOI: 10.1557/opl.2012.765

Au and NiO nanoparticles dispersed inside porous SiO₂ sol-gel film: correlation between localized surface plasmon resonance and structure upon thermal annealing

Enrico Della Gaspera[1], Giovanni Mattei[2] and Alessandro Martucci*[1]

[1]Università di Padova, Dipartimento di Ingegneria Industriale, Padova, Italy
[2]Università di Padova, Dipartimento di Fisica e Astronomia, Padova, Italy

ABSTRACT

The favorable lattice matching between Au and NiO crystals made possible the growth of unique cookie-like nanoparticles (25 nm mean diameter) inside a porous SiO₂ film after annealing at 700 °C. The unusual aggregates result from the coupling of well distinguishable Au and NiO hemispheres, which respectively face each other through the (100) and (200) lattice planes. The thermal evolution of the Au and NiO nanoparticles structure has been studied by high resolution transmission electron microscopy and UV-visible absorption spectroscopy and correlated with the evolution of the Au surface plasmon resonance peak.

INTRODUCTION

Films containing NiO, a p-type semiconductor with a wide band gap of 4.2 eV, have been recently proposed as sensitive materials for chemoresistive[1] and optical gas sensors[2]. The working mechanism of NiO-based platforms consists in a change of the electrical resistance or optical transmittance of the material as a consequence of physisorption/chemisorption and reactions of the analyte with its surface. The interaction extent with the target gas can be maximized either by increasing the sensor's surface area[3], or by doping the oxide with noble metal nanoparticles (NPs)[4]. In previous works we demonstrated that SiO₂ porous films containing NiO NPs have remarkable gas sensing properties[2,5]; moreover the optical gas sensing effect can be enhanced by the introduction of Au NPs together with the NiO nanocrystals[6,7]. In fact, the surface plasmon resonance (SPR) of Au NPs depends strongly upon the surrounding medium properties, particularly refractive index[8]. Even a small change of the matrix properties leads to a large variation of the plasmonic frequencies, allowing for detection of the target gas at low concentrations. Moreover, thanks to the Au SPR peak in the visible region, it is possible to tune the response of the sensors, by selecting an appropriate wavelength of analysis[9,10]. In fact, the variation of the dielectric constant around the SPR bands will differ for different gas species; this leads to a diverse variation of the optical properties at different wavelengths. In this way it is possible to enhance the selectivity towards interfering gases, a feature that is not achievable with conventional electrical sensors

We already have shown that by tailoring the annealing conditions it is possible to change the NP morphology inside the porous SiO₂ matrix[11]. At 500 °C the NiO and Au NPs are separated while at 700 °C the clusters show a twofold structure formed by two hemispheres of NiO and Au arranged in a cookie-like structure, with the Au (111) planes parallel to NiO (200) ones. The objective of this work is to study the thermal evolution of the structure of the SiO₂ film containing Au and NiO NPs upon annealing from room temperature up to 1000 °C and to correlate the evolution the SPR peak with the structural changes of the Au and NiO NPs.

EXPERIMENT

The nanocomposite films were prepared by the sol-gel method mixing a matrix solution containing the precursors for silica and gold, and a doping solution containing the precursor for NiO. Detailed experimental procedures have been reported elsewhere[6,9]. The precursor solution for the silica matrix was made by mixing tetra-ethoxy silane (TEOS), methyl-triethoxy silane (MTES), H_2O, HCl in EtOH. The gold precursor ($HAuCl_4*3H_2O$) was added directly in the matrix solution. The NiO precursor solution was prepared mixing $NiCl_2*6H_2O$ in EtOH in the presence of N-[3-(trimethoxysilyl)propyl]-ethylenediamine (DAEPTMS). Nominal molar ratios $SiO_2:NiO=7:3$ and Ni:Au=5:1 have been used. The two precursor sols were then mixed together to obtain the final batch used for film deposition. Films were deposited on silicon or silica glass substrates via the dip-coating technique and annealed progressively from 100 °C to 1000 °C in air, with 30 minutes steps every 100 °C.

Transmission Electron Microscope (TEM) analyses on cross-sectional samples of the composite films were taken with a field-emission FEI TECNAI F20 SuperTwin FEG instrument operating at 200 kV and equipped with an EDX energy-dispersive x-ray spectrometer (EDS) for compositional analysis. Scanning TEM (STEM) analysis coupled with EDS allowed compositional analysis on single clusters by means of line scans with an electron probe resolution of 1 nm FWHM.

X-ray diffraction (XRD) patterns were recorded using a Philips PW1710 diffractometer equipped with glancing-incidence X-Ray optics. The analysis was performed at 0.5° incidence using CuKα Ni-filtered radiation at 30 kV and 40 mA.

Optical absorption spectra of the films deposited on silica glass were measured in the 300-800 nm range using a Jasco V-570 spectrophotometer.

RESULTS AND DISCUSSION

In figure 1 the absorbance spectra of the prepared films are reported. The plots confirm the Au NPs formation after annealing at 200 °C, evidenced by the pronounced SPR band centered at 540 nm. The 3D plot allows to appreciate an increase in the SPR intensity in the 200 °C – 500 °C annealing interval, with a maximum value after treatment at 500 °C. A slight broadening of the band is also appreciable, although its shape clearly remains that of a single peak centered in the same spectral region. This trend can be explained by Au NPs progressive growth during annealing. In a previous work we have demonstrated that at 500 °C the nanocomposite films consist of separated NiO (about 5 nm diameter) and Au (about 25 nm diameter) NPs immersed in a porous SiO_2 matrix[11].

At higher temperatures, between 600 °C and 800 °C, the band showily broadens and decreases in intensity, with appearing of the two-featured structure after treatment at 700 °C. We have demonstrated that this double shaped band is due to the formation of a twofold structure formed by two hemispheres of NiO and Au arranged in a cookie-like structure, with the Au (111) planes parallel to NiO (200)[7]. An example of such double structure is reported in figure 2. We observed the formation of cookie-like structures even at 600 °C but for longer annealing times (at least 4 hours isothermal treatment), whilst no twofold clusters are detected at lower temperatures even for very long annealing (up to 20 hours). Surprisingly, in the 900 °C – 1000 °C region the band shows a clear blue-shift, in opposition to the trend observed at lower temperatures, maintaining its double featured shape at 900 °C (even if the difference in wavelength between the two peaks is strongly reduced) but recovering a much more pronounced lower wavelengths single contribution at 1000 °C. These features suggest a decomposition of the Au-NiO cookie-like structure with possible dissolution of the NiO and hence removal of the Au/NiO interface which is responsible of the double SPR band[7].

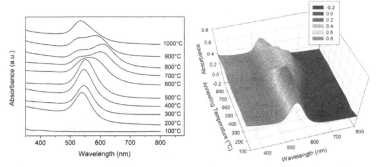

Figure 1 Optical absorption spectra of SiO_2:NiO:Au film annealed from 100 °C to 1000 °C. Left: vertically translated spectra to better appreciate the SPR band thermal evolution. Right: tri-dimensional plot of spectra showing the actual evolution of the band intensity (evidenced by color fading) and shape.

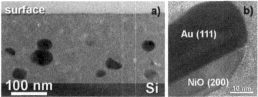

Figure 2 a) Cross section bright field TEM image of a film annealed up to 700 °C. b) High resolution TEM image of a cookie-like cluster.

These assumptions have been supported by XRD analyses (see figure 3). For the sake of clarity, only the diffraction patterns for the films annealed from 500 °C to 1000 °C have been reported, since at lower temperatures NiO is amorphous. Gold clusters grow progressively with increasing the annealing temperature, even if the crystallite size is smaller than the actual NPs size observed from TEM images. This can be explained with Au NPs being polycrystalline. As far as NiO is concerned, at 500 °C no diffraction peaks have been detected, possibly because of the very small size of NiO clusters and their poor crystallinity. Nevertheless previous TEM analyses do show the presence of small NiO crystals after 500 °C annealing[11]. At higher temperatures NiO diffraction peaks start to appear and they become sharper with increasing the annealing temperature, suggesting NPs growth. The crystallite size estimated with the Scherrer relationship from the FWHM of the diffraction peaks shows a maximum at 800 °C, consistent with the greatest red shift of the SPR band observed in the optical absorption spectra reported in figure 1. Starting at 900 °C, NiO crystallite size decreases, confirming the decomposition of the cookie-like structures inferred from the optical spectra.

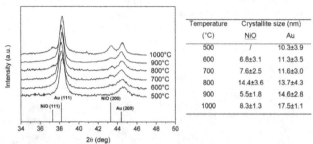

Temperature	Crystallite size (nm)	
(°C)	NiO	Au
500	/	10.3±3.9
600	6.8±3.1	11.3±3.5
700	7.6±2.5	11.6±3.0
800	14.4±3.6	13.7±4.3
900	5.5±1.8	14.6±2.8
1000	8.3±1.3	17.5±1.1

Figure 3 XRD patterns of the films annealed from 500 °C to 1000 °C. The crystallite size estimated with the Scherrer relationship for both NiO and Au is reported in the table.

To better clarify the morphologic cause of this behavior, the film annealed up to 1000 °C was analyzed by TEM. A cross sectional TEM image of such sample is shown in figure 4a. Estimated film thickness was 133 ±10 nm. Huge Au NPs embedded in the silica network are clearly visible, giving high contrast in the image, while small and clearer NiO NPs are finely dispersed all through the film thickness. Au NPs appear to be spherical in shape, so a question arises about the eventual evolution of the two-folded structures observed in 700 °C annealed samples. Optical spectra seem to indicate that the influence of the dielectric discontinuity is drastically reduced, so it must be postulated that somehow NiO re-dissolved inside the SiO_2 matrix in form of fine NPs , as also confirmed by the crystallite size evaluated from XRD, although a minimum amount of it still remains coupled on the Au surface, as testified by both a closer view of the TEM image and by presence in the SPR band of a high wavelength absorbance tail. Closer views of the film section, as in figure 4 b and c, permit to appreciate the spherical shape of the Au NPs, that barely present themselves interfacing the NiO lattice. Size of the two NPs families was estimated from lower magnification images and results are plotted in figure 4 d and e.

The big spherical Au aggregates have an average diameter of 49 nm with standard deviation of 13 nm, while the size of the small NiO clusters was estimated to be 3.2 ± 0.7 nm. The nature of the aggregates has been confirmed by EDS compositional analysis, performed on the highlighted areas in TEM image of figure 4b and summarized in the table of figure 4. The beam focused on the higher contrast areas (zone A) confirmed that the big spherical aggregates are extremely rich in Au, while signals related to Ni atoms is lower. Presence of Si and O is due to the silica walls crossed by the incident beam that are surrounding the Au aggregates out of the image planes. Ni signal may raise from those portions as well and not from inside the Au aggregates. In the rest of the matrix (zone B) Au is no more detectable and only signals raising from Ni, Si and O are evident. Atomic ratio between O and the other two species indicates that a certain fraction is involved in the SiO_2 structure and the remaining in NiO. The clear 35 nm thick layer visible between the film and the silicon substrate (zone C) was found to be pure SiO_2, most likely originated by oxidation of the silicon substrate during thermal annealing at 1000 °C.

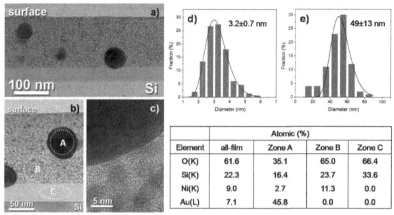

	Atomic (%)			
Element	all-film	Zone A	Zone B	Zone C
O(K)	61.6	35.1	65.0	66.4
Si(K)	22.3	16.4	23.7	33.6
Ni(K)	9.0	2.7	11.3	0.0
Au(L)	7.1	45.8	0.0	0.0

Figure 4 Bright field TEM analysis of a SiO₂:NiO:Au film annealed progressively up to 1000 °C. a) Low magnification cross sectional image of the nanocomposite film. b) Higher magnification image of the same film: dotted circles indicate areas in which EDX analysis was performed. EDX results are summarized in the table. c) High resolution image showing a portion of a big Au NP and several small NiO clusters. The histograms of the size distribution for NiO and Au clusters calculated from TEM images are reported in (d) and (e) respectively.

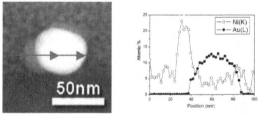

Figure 5 Left: STEM image of an Au cluster. Brighter area is related to higher contrast material, Au. The arrow indicates the direction of the electron beam during scan used to perform EDX analysis. Right: EDX profiles resulting by moving the electron beam across the cluster.

EDX analysis was performed moving the electron beam across one of the big clusters, and obtained plots of atomic profiles are reported in figure 5. The analysis thus confirmed what suggested by the absorbance spectra. After annealing at 1000 °C all the Au is involved in big spherical clusters, with no trace of it remaining inside the rest of the matrix core. Cookie-like structures as seen in the 700 °C annealed sample disappear, and NiO seems to re-dissolve inside the SiO₂ matrix in form of fine nanoparticles. EDX scan of figure 5 excludes the complete dissolution of NiO that is still partially coupled with small portions of the Au NPs surface, therefore explaining the absorbance tail still present in the absorbance spectra in the 550 - 600 nm wavelength range

CONCLUSIONS

Formation and decomposition of cookie-like structures composed of Au and NiO hemispheres inside a sol-gel SiO_2 matrix has been analyzed with optical spectroscopy coupled to TEM and XRD morphological and structural investigation. At 500 °C Au and NiO clusters are separated, while starting at 600 °C two-fold structures appear, resulting in a distinctive optical absorption behavior. These cookie-like clusters are stable up to 800 °C, but at higher temperatures NiO NPs seem to re-dissolve inside the silica matrix, causing a drastic blue shift and a change in the shape of the Au SPR band. At 1000 °C, big Au NPs and very small NiO NPs are almost completely separated inside the silica matrix.

REFERENCES

1. I. Hotovy, J. Huran, P. Siciliano, S. Capone, L. Spiess and V. Rehacek, *Sens. Act. B* **103,** 300 (2004).
2. A. Martucci, M. Pasquale, M. Guglielmi, M. Post and J.C. Pivin, *J. Am. Cer. Soc.* **86,** 1638 (2003).
3. X.Wang, G. Sakai, K. Shimanoe, N. Miura and N. Yamazoe, *Sens. Act. B* **45,** 141 (1997).
4. M. Matsumiya, W. Shin, N. Izu and N. Murayama, *Sens. Act. B* **93,** 309 (2003).
5. A. Martucci, D. Buso, M. De Monte, M. Guglielmi, C. Cantalini and C. Sada, *J. Mat. Chem.* **14,** 2889 (2004).
6. D. Buso, M. Guglielmi, A. Martucci, G. Mattei, P. Mazzoldi, C. Sada and M. L. Post, *Nanotechnology* **17,** 2429 (2006).
7. G. Mattei, P. Mazzoldi, M. L. Post, D. Buso, M. Guglielmi and A. Martucci, *Adv. Mater.* **19,** 561 (2007).
8. P.H. Rogers, G.Sirinakis and M.A. Carpenter, *J. Phys. Chem. C*, **112,** 6749 (2008).
9. D. Buso, G. Busato, M. Guglielmi, A. Martucci, V. Bello, G. Mattei, P. Mazzoldi and M. L. Post, *Nanotechnology* **18,** 475505 (2007).
10. M. Ando, T. Kobayashi and M. Haruta, *Catalysis Today* **36,** 135 (1997).
11. D. Buso, M. Guglielmi, A. Martucci, Giovanni Mattei, P. Mazzoldi, C. Sada and M.L. Post, *Cryst. Growth & Design* **8,** 744 (2008).

Thin Films, Ceramics, Nanoparticles, and Applications

Mater. Res. Soc. Symp. Proc. Vol. 1449 © 2012 Materials Research Society
DOI: 10.1557/opl.2012.794

Ferromagnetism in Nanocrystalline Powders and Thin Films of Cobalt-Vanadium co-doped Zinc Oxide

Marco Gálvez-Saldaña[1], Gina Montes-Albino[2] and Oscar Perales-Perez[3]

[1]Department of Physics, University of Puerto Rico at Mayagüez, Mayagüez 00980, PR, USA.
[2]Department of Mechanical Engineering, University of Puerto Rico at Mayagüez P.O. Box 9045, Mayagüez, PR, 00681-9045 USA.
[3]Department of Engineering Science and Materials, University of Puerto Rico at Mayagüez, Mayaguez, PR, 00680-9044, USA.

ABSTRACT

A systematic study was carried out to determine the effect of the composition and annealing atmosphere (air and N_2) on the structural, optical and magnetic properties of pure, doped and co-doped ZnO [$Zn_{(1-y)}(CoV)_yO$] nanocrystalline powders and films. The (Co+V) doping level, 'y', was fixed at 2 at% with variable individual concentrations of Co and V species. Powders and films were synthesized via a sol-gel approach where the films were grown on silicon (100) substrates. X-ray diffractometry verified the formation of the ZnO host structure after annealing of the precursor phases. The variation of the average crystallite size of Co-V (2 at.%) ZnO powders annealed in air at 500°C were negligible and averaged 33 nm. Photoluminescence (PL) measurements of powder corroborated the formation of high-quality ZnO host structure, as well as in films annealed in air. In turn, XRD and PL measurements confirmed an enhanced crystallinity of the ZnO host, with an average crystallite size of 41 nm, for films annealed at 500°C under a N_2 atmosphere. M-H measurements evidenced a ferromagnetic behavior at room temperature in powders and films that was dependent on the type and amount of the dopant species.

I. INTRODUCTION

ZnO is a II-VI semiconductor type with hexagonal wurtzite structure and belongs to the space group C_{6V}^4 or $P6_3mc$. The corresponding hexagonal unit cell has two lattice parameters: 'a'= 3.249Å and 'c'=5.207Å [1]. The larger exciton binding energy ~60 meV and wide band gap ~ 3.3 eV at room temperature increase the technological attractiveness of this oxide. In turn, the effective incorporation of dopant species into ZnO host structure usually induces changes in the physical and chemical structure, leading to novel multi-functional properties. In the case of ZnO doped with transition metal ions, the subsequent exchange interaction between available spins of the magnetic species would induce a ferromagnetic behavior in the so-called ZnO-based diluted magnetic semiconductor (DMS) [2,3]. This ferromagnetic functionality would enable the application of this material in data storage and spintronics-based devices. The rationale behind the synthesis of co-doped (Co + V) ZnO relies on the expected synergistic effect of these doping species, based on our earlier work where doping of ZnO with Co or V induced a strong room-temperature ferromagnetism [4]. In addition, there is a lack of systematic studies regarding the effect of co-doping of ZnO on the corresponding optical and magnetic properties in thin films.

There are numerous deposition techniques to synthesize ZnO thin films, including chemical vapor deposition, sputtering, pulsed laser deposition and sol-gel [5-8]. The sol-gel approach was selected because better control on the chemical composition in the developed oxide structures can be achieved by this route. Accordingly, powders and films of ZnO were simultaneously doped with V^{3+} and Co^{2+} ions. The structural, optical and magnetic properties of the produced materials were studied as a function of the dopant levels and the annealing atmosphere (air or nitrogen).

II. EXPERIMENTAL

A. Materials
Zinc acetate [Zn (CH_3COO) $2H_2O$, purity 98%], cobalt acetate [Co $(CH_3COO)_2$ $4H_2O$, purity 99%] and vanadium (III) chloride [VCl_3, purity 97%] were used as precursor salts. The (Co+V) doping level, 'y' in [$Zn_{(1-y)}(CoV)_yO$], was fixed at 2 at% with variable individual concentrations of Co and V species. Ethanol and monoethanolamine were used as the solvent and the viscosity-controlling additive, respectively. The total concentration of metals ions in precursor solutions was kept constant at 0.2mol/L in all experiments. Silicon substrates, purchased from Addison Engineering Inc., were thoroughly cleaned before spin-coating.

B. Synthesis of Powders and Thin Films
Powders were synthesized by dissolving appropriate amounts of Zn, Co and V(III) salts in ethanol, followed by their heating at 120°C for 12 hours to guarantee the complete removal of the solvent. The obtained solid precursor was then annealed in air for 1 hour at 500°C to obtain the desired crystalline structure. The synthesis of thin films began by dissolving the precursor salts in ethanol. Then, monoethanolamine was added to 10 mL of the metal precursor solution to control viscosity. Next the solution was ultrasonicated for 5 minutes to obtain a clear solution that was added drop-wise onto a clean Si (100) substrate and spin-coated at 3000 rpm for 20 seconds. After each coating cycle, produced films were pre-dried at 50°C and 200°C for 3 minutes to remove organic residuals. These spin-coating/drying cycles were repeated twenty times to thicken the films up to 1000 nm, approximately. Finally, the films were annealed in air and nitrogen atmosphere for one hour at 500°C. The heating rate was 10°C per minute for each sample.

C. Materials Characterization
The structures of the synthesized powders and films were determined by using a SIEMENS D500 X-ray diffractometer (XRD) with Cu-Kα radiation. The optical properties of the powders and films were measured using a UV-vis DU 800 spectrophotometer and a RF 5301 PC Spectrofluorophotometer Shimadzu. The excitation wavelength for the photoluminescence (PL) measurements was 342 nm. The magnetic hysteresis loops (MH), were recorded at room temperature using a Lake Shore 7410 VSM magnetometer.

III. RESULTS AND DISCUSSION

A. X-ray Diffraction Analyses
The X-ray diffractograms in Figure 1 correspond to the (V-Co) ZnO powders annealed in air for one hour at 500°C. The samples were synthesized keeping constant the total dopant concentration (2 at.%) for the Co-doped, V-doped and (Co+V) co-doped ZnO powders. All diffractions peaks are assigned to the host ZnO phase with a hexagonal wurtzite structure; no secondary phases of vanadium or cobalt oxides were detected, which suggests the actual incorporation of the dopants in the ZnO lattice. The actual incorporation of the dopants into the

ZnO lattice could be expected because of the similar ionic ratios of Co^{2+}(0.72Å), V^{3+}(0.78Å) and Zn^{2+} (0.74Å). The incorporation of V^{3+} ions into the ZnO host lattice was conducive to the decrease in the average crystallite size from 30 nm, in pure ZnO, to 15 nm in V(2 at.%)-ZnO; this effect could be attributed to the inhibiting effect of vanadium ions on the oxide crystallization during the thermal decomposition process. In the case of Co- and (Co+V)-ZnO, the presence of dopants did not produce changes in the average crystallite size with respect to the pure material. The corresponding lattice parameters were estimated at "a"= 3.247 Å and "c"= 5.206 Å. Figure 2 shows the diffraction patterns of pure and (V^{3+}+ Co^{2+}) ZnO films annealed at 500°C in air, (a), and nitrogen, (b). Similarly, all diffraction peaks are assigned to the ZnO hexagonal wurtzite phase. In this case, the co-existence of V and Co dopants caused a decrease of the average crystallite size in the films from 31 nm, in pure ZnO, to 16 nm in the (Co+V)ZnO sample annealed in air. In the case of films annealed in N_2 atmosphere, the average crystallite size was around 41 nm in all samples. The XRD patterns of Figure 2 also evidence the preferential growth of the (002) plane; the lower surface energy attributed to this plane can explain this trend [9]. This preferential growth was enhanced even further when the films were annealed in N_2 atmosphere. The annealing in N_2 also enhanced the sample crystallinity, which could also explain the preferential growth of the (002) plane.

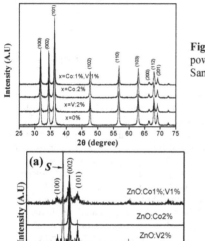

Fig.1.- XRD patterns of pure and doped ZnO powders with fixed 2 at.% dopant concentration. Samples were annealed for one hour at 500°C.

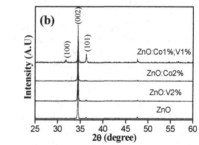

Fig.2.- (a) XRD diffractograms of pure and doped ZnO films deposited onto Si (100) substrates annealed (a) in air and (b) in nitrogen atmosphere. Samples were annealed for one hour at 500°C. The peak *S* corresponds to the Si substrate.

B. UV-vis Measurements

Figure 3 shows the UV-vis spectra of pure and (Co+V) doped ZnO/quartz films annealed at 500°C in air. The slight but noticeable blue-shift in the absorption peak can be attributed to the actual incorporation of V^{3+} and Co^{2+} species in the ZnO lattice; dopants can induce the generation of superficial defects causing changes in the optical properties [10]. This blue-shift can also be explained in terms of the Moss-Burstein effect, which takes place when the carrier concentration exceeds the conduction band due to doping. As the dopant concentration is increased, more donor states are generated which places the Fermi level inside the conduction band, generating an apparent increase in the band gap [11].The band gap energy for pure and doped films was estimated using Tauc's relationship [12] and were in the 3.27-3.30 eV range.

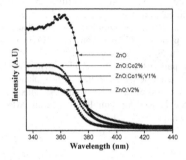

Fig. 3.- UV-Vis of (Co+V) ZnO/quartz films and annealed for one hour in air at 500°C.

C. Photoluminescence Measurements

Figure 4 shows the room-temperature PL spectra of pure and doped ZnO powders annealed for one hour in air at 500°C. The spectra were recorded using an excitation wavelength of 342 nm. As seen, the main emission band was centered on 388 nm; no bands in the visible region were observed in powdered samples. This fact may suggest that the selected synthesis route did not favor the creation of defects which are usually associated with the luminescence in the visible region. In turn, the drop in the intensity of the emission band with doping suggests a quenching-by-dopant effect. This type of quenching has been observed in other systems when the concentration of the dopant becomes so high that the probability of non-radiative transition through trapping states exceeds the probability of radiative transition. This in turn causes the excitation energy to start migrating through lattice vibrations [13]. Figure 5 shows the PL spectra of pure and doped ZnO/Si (100) thin films annealed at 500°C, in air (a), and nitrogen, (b). The spectra were recorded at the same wavelength as for the powdered samples. A strong emission band in the UV region was observed in all samples. The fact that doped films annealed in air did not exhibit emission in the visible region may suggest that dopant species occupied the Zn vacancies, usually responsible for this emission. Also in this case, a quenching effect in the UV emission band was observed, particularly when ZnO was doped with Co species. Figure 5-b shows the PL spectra for the films annealed in nitrogen. Again, quenching of the UV luminescence bands was observed; however, the drop in the band intensity was not as pronounced as in those films annealed in air. The high crystallinity of the samples annealed in nitrogen could be related to the observed inhibition of the quenching in luminescence. On the other hand, a strong visible emission band was observed only in pure and V (2 at. %)- ZnO films. This band is usually attributed to lattice defects like oxygen vacancies [14-15]. Recent theoretical works attribute this band to zinc vacancies that posses lower formation energy than oxygen vacancies, or zinc interstitials [16]. Moreover, the annealing under inert conditions provided by N_2 should have favored the formation of these lattice defects, as observed in V-doped ZnO [17].

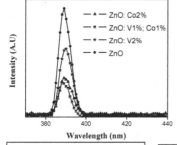

Fig.4.- Room-temperature PL spectra for pure and doped ZnO powders, annealed in air at 500°C.

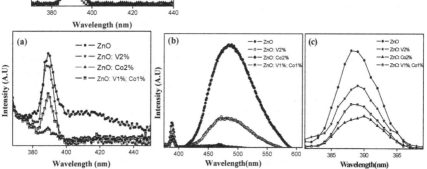

Fig. 5.- Room-temperature PL of pure and doped ZnO/Si(100) films and annealed at 500°C for one hour in air, (a), and nitrogen, (b and c). Figure (c) shows the PL information in the UV region.

D. Magnetic Measurements

Fig. 6 shows room-temperature M-H loops of V- and (Co+V)-doped ZnO/Si (100) films, annealed at 500°C in air, (a), and nitrogen, (b). A weak but noticeable ferromagnetism was observed in all the samples.

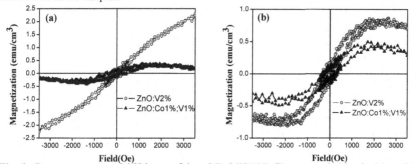

Fig. 6.- Room-temperature M-H loops of doped ZnO/Si(100) films, annealed in air, (a), and N₂, (b)

The coercivity was increased from 225 Oe to 270 Oe for (Co+V) ZnO films annealed in air and nitrogen, respectively. This ferromagnetic behavior may be attributed to the super exchange interaction through oxygen ions (TM-O^{2-}-TM, where TM: transition metal ion) or to the exchange interaction between spins of the band carriers and localized spins of Co and V. Cobalt ions contribute to the saturation magnetization with $6\mu_B$ and Vanadium ions with $4/3\mu_B$ per ion. The films annealed under nitrogen atmosphere (Figure 6-b) exhibited a stronger ferromagnetic response that can be attributed to their enhanced crystallinity, as suggested by XRD measurements. Moreover, the observed ferromagnetic response can also be related to the generation of structural defects favored under a nitrogen atmosphere, in agreement with the information provided by PL analyses (Figure 5). This defect-induced ferromagnetism can be explained in terms of the bound magnetic polarons model (BMP) applied to oxide DMS's [18].

CONCLUSIONS

The formation of pure, doped and (V^{3+}+Co^{2+}) co-doped ZnO nanocrystalline powders and thin films were confirmed from a structural and optical standpoint. In addition, important results were obtained in the magnetic properties, coexisting with the semiconductor intrinsic properties at room temperature for the films annealed under nitrogen. These properties convert this ZnO based DMS in a multifunctional material with novel applications in optoelectronic, data storage devices and spintronics.

Acknowledgments
This material is based upon work supported by the DOE-Grant No FG02-08ER46526.
Special thanks to M.S. Boris Rentería, UPRM, for the magnetic measurements at the **NANO** materials Processing Laboratory at UPRM

References
[1] Ü. Özgür, et al., Journal of Applied Physics **98**, 041301, (2005).
[2] F. Pan, C. Song, et. al, Materials Science and Engineering R **62**, 1-35, (2008).
[3] Jing Qi, Daqiang Gao, et. al, Appl Phys A. **100**, 79-82 (2010).
[4] M. Gálvez, et.al, Mater. Res. Soc. Symp. Proc. Vol. 1368, DOI: 10.1557/opl.2011.1038.
[5] T.M. Barnes, et.al, J. Cryst. Growth, **274**, 412-417, (2005).
[6] D.J. Kang, et. al, Thin Solid Films **475,** 160– 165, (2005).
[7] J. Hyun Kim,et. al, Journal Applied Physics, **92**, 10, (2002).
[8] J. Petersen, Microelectronics Journal **40,** 239–241, (2009).
[9] Seong Keun Kim, et. al, Thin Solid Films **478**, 103– 108, (2005)
[10] Liwei Wang, et. al, Thin Solid Films **517,** 3721–3725 (2009)
[11] Marius Grundmann, Handbook *The Physics of Semiconductors: An Introduction Including Nanophysics and Applications*, Springer, Second edition, 291-292, (2010).
[12] J.J. Lu, et. al, Optical Materials **29**, 1548–1552, (2007).
[13] S. Shionoya and W.M. Yen, *Phosphor Handbook*, CRC Press, Boca Raton, Florida (1999).
[14] K. Vanheusden, et. al, Appl. Phys. Lett. **68**, 403, (1996).
[15] P. K. Samanta, International Journal of NanoScience and Nanotechnology ISSN 0974– 3081, **1**, 81-90 (2009).
[16] Anderson Janotti and Chris G Van de Walle, Rep. Prog. Phys.**72**, 126501, (2009).
[17] Shubra Singh, et. al, New Journal of Physics **12**, 023007 (2010)
[18] J.M.D. Coey, M. Venkatesan, C.B. Fitzgerald, Nat. Mater. 4, 173, (2005).

Mater. Res. Soc. Symp. Proc. Vol. 1449 © 2012 Materials Research Society
DOI: 10.1557/opl.2012.959

Modification of Cordierite Honeycomb Ceramics Matrix for DeNOx Catalyst

Qingcai Liu, Yuanyuan He, Jian Yang, Wenchang Xi, JuanWen, Huimin Zheng
College of Materials Science and Engineering, Chongqing University, Chongqing, 400030,
China

ABSTRACT

To obtain highly dispersed and highly active catalysts by impregnating of active species onto the monolith directly, cordierite honeycomb ceramics were modified by nitric acid solution of 68wt%. Effects of acid treatment temperature and time on the performance of cordierite were investigated. Specific surface area, pore size distribution, morphology and structure of cordierite were characterized by N_2-physical adsorption, SEM, XRD, respectively. Concentrations of ions in the acid solution were measured by AAS. It is shown that the corrosion content of cordierite increases and more micropores are generated with increasing time of acid treatment, leading to an upward trend of specific surface area. The coefficient of thermal expansion and compression strength decrease obviously at a higher temperature, which is mainly attributed to the removal of Al and Mg ions from the silicate structure and delayed formation of free amorphous silica on the surface of the cordierite. The optimal modification process of cordierite matrix acid erosion is at 110℃ for 6 h.

INTRODUCTION

Cordierite($2MgO \cdot 2Al_2O_3 \cdot 5SiO_2$) is a crystalline magnesium alumosilicate with hexagonal framework. It is the most widespread commercial material for high-temperature catalyst applications, especially the substrate for automobile exhaust catalyst, because the cordierite has some special characteristics such as high mechanical stability and low coefficient of thermal expansion (CTE). Cordierite is in the form of a honeycomb monolith because of the requirements of low pressure drop and high geometrical surface area. However, cordierite monoliths are not suitable for using as catalyst supports due to quite small specific surface area. It is not possible to obtain highly dispersed and highly active catalysts by impregnating of active species onto the monolith directly[1,2].

It has been reported that alumina, perovskites, silica, zirconia, titania, and zeolites are effective to perform the washcoating, and lead to high BET surface areas by depositing a layer of these highly porous materials[3-5]. In addition to the washcoating process, organic or inorganic acid treatment on the surface of cordierite monoliths has been studied and proves to be an available approach to increase surface areas[5-9]. Furthermore, the relevant studies focus on the select of acid and the processing conditions[6,8-10]. Whereas the influence of acid leaching on the CTE and mechanical strength of cordierite honeycomb ceramics were rarely reported from previous studies[2,10]. In the current work, therefore, the CTE and mechanical strength of cordierite were discussed to optimize the time and temperature for nitric acid (68% of concentration) treatment. Specific surface area, pore size distribution, morphology and structure of acid-corrosion cordierites were characterized by N_2-physical adsorption, SEM, XRD. Concentrations of ions in the acid treatment solution were measured by AAS. In particular, the optimal modification process for cordierite matrix acid erosion was obtained.

EXPERIMENTAL

Nitric acid treatment of Cordierite

Cordierite honeycomb ceramic (Jiangsu yixing non-metallic chemical machinery Co. Ltd, having 400 cells per square inch(cpsi)) was used for all experiments. Cordierite core samples with a length of 10 mm, diameter of 10 mm and height of 20 mm were cut from the cordierite monoliths, afterwards ultrasonic cleaned, dry for chemical treatments and detections. The samples were then immersed in nitric acid solution of 68 wt% at 20℃ and 110℃ for 3h, 6h, 9h and 12h, respectively. Thereafter the resulting solids were thoroughly washed with distilled water to neutrality. Finally, the solids were kept at 110℃ for 2 hours to be dried.

Characterization

The coefficient of thermal expansion (CTE) was measured by a Netzsch DIL402C thermal dilatometer with an ascending temperature rate of 10℃/min from room temperature to 1000℃. Compression strength was determined by a WDW-200B universal testing machine. It was then calculated according to the following expression. $\sigma = \dfrac{P}{A}$, where σ is the compression strength (Pa), P the fracture load (N), and A is the load area of samples(m^2). In addition, weight losses resulted from the treatments were calculated simply by weighing samples before and after acid treatment. Textural characterization, specific surface areas, and pore size distributions of samples were determined by nitrogen adsorption-desorption isotherms at 77K (Micromeritics ASAP 2010 apparatus) after 2 h pretreating under vacuum at 300℃. Scanning electron microscopy was used to observe the morphological changes produced by the acid treatment. XRD (Rigaku D/max-1200 X-ray diffractometer with Cu K α radiation) was employed in the present study with the step scans were taken over the range of 2θ from 5° to 85° at a scanning rate of 5°/min. The concentrations of the ions dissolved in the treatment solutions were measured by Z-8000 atomic absorption spectroscopy.

RESULTS AND DISCUSSION

Influence of acid treatment on physical properties of cordierite matrix

Figure 1 shows results of CTE of cordierite after acid treatment. It can be seen clearly that the CTE decrease gradually with extension of acid treatment time which is similar to results described in the previous studies[2,10]. An approximately linear relationship exists between CTE and treatment time for both series samples that treated at different temperatures. However, the cordierite samples which experienced acid treatment at 110℃ have much lower CTE than the samples treatment at 20℃. Nitric acid effectively improves the stability of high temperature especially acid corrosion at 110℃. This tendency can be explained by that destruction reaction of the silicate structure of cordierite is more drastic at a high temperature, with the removal of Mg and Al ions and generation of free silica on the surface of the cordierite. According to previous reports[1,2], the free silica left in the acid-treated cordierite, either in crystal line or in amorphous form, has a tremendous influence on the thermal expansion properties of cordierite ceramics acid-treated. This also can be confirmed by XRD spectrum diagram shown below.

Figure 2 displays that axial compressive strength of samples shift downward more rapidly at a higher temperature. After acid corrosion at 110℃, the axial compressive strength are 25.8MPa for 6 h and 19.2MPa for 12 h, respectively. It also suggests that the cordierite matrix dissolved into the corrosion medium partly and therefore a structure of more porous was formed. After treatment for a long time, the porous cordierite samples become fragile because the fracture strength decreases greatly at high acid leaching levels. It is suitable for cordierite matrix corroding in 6 h to ensure sufficient mechanical strength.

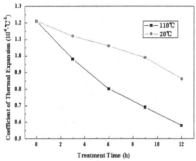 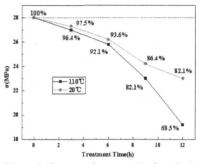

Figure 1. The CTE of cordierite samples   **Figure 2.** Compressive strength of cordierites

The weight loss rate is shown as a function of treatment time in Figure 3. Compared with low temperature, cordierites treatment at 110℃ show a faster acid-corrosion speed. For the same corrosion time, the mass loss is far greater than that of acid corrosion at 20℃. With increasing treatment time, the mass loss of all samples increase slightly at low temperature while increase drastically at higher temperature especially after 6 h later. The weight loss of cordierite sample acid-etched for 12 h is found to be as high as 6%. The rapid acid-corrosion rate of the cordierite is mainly due to the nature of cordierite, as well as the existence of alkaline metal and alkaline earth metal oxides in it. This corresponds to the removal of the Al, Mg, Ca and Na ions, etc from the cordierite lattice, as confirmed by AAS data shown below. After all, the cordierite shows a poorer acid-corrosion resistance under a higher temperature. From these data it can be concluded that acid-corrosion at 110℃ is appropriate for cordierite in terms of efficient.

BET specific surface area and pore size distribution variations

The specific surface area, pore size and pore volume of cordierites treatment at 110℃ are listed in Table I . There is a significant increase in BET surface area after acid treatment. After acid corrosion for 12 h at 110℃, the specific surface area is found to increase greatly under these conditions to value as high as 41.95 m^2/g, over 33 times higher than that of the original cordierite monolith. These results suggest that acid corrosion under these conditions could obviously improve the surface area of matrix. Similarly, pore volume increases significantly after acid treatment because of the dissolution of MgO and Al_2O_3 and thus the creation of micropores and mesopores. Nevertheless, the pore volume of sample eroded after 9 h is downward comparing with 6 h and the pattern in average pore diameter is rather different.

Table I . Surface area, pore size and pore volume of various samples treatment at 110℃

Treat time	0	3	6	9	12
$S_{BET}(m^2 \cdot g^{-1})$	1.25	7.88	14.65	23.69	41.95
$V_{pore}(cm^3 \cdot g^{-1})$	0.00131	0.00511	0.02460	0.01524	0.03250
$D_{pore}(nm)(BET)$	4.2	2.6	6.72	2.57	3.1

Pore size distribution of samples is presented in Figure 4. It can be inferred (both figure4 and table 1) that the increase of surface area is accompanied by creation of micropores and mesopores centered between 1 and 3 nm diameter, whereas the original cordierite contains a small amount of pores predominantly between 2 and 3 nm and consequently shows a small surface area of 1.25 m^2/g and a small pore volume of 0.00131 cm^3/g. The micropores may be generated by direct removal of metal ions from the surface of cordierite, but some rearrangement of the structure appears necessary in order to produce mesopores. A redeposition of dissolved silica may be involved in this process[1,2]. The cordierite treatment for 6h obtains a lot of pores between 1 and 10 nm, a wide range of pore size distribution. Moreover, it shows a bimodal distribution of pore size with two peaks appearing around 1.5nm and 6nm respectively, which is the most optimal pore structure for DeNOx catalyst matrix. Micropores around 1.5nm may be new formed as a result of acid corrosion while mesopores around 6nm may be due to growing of original pores between 2 and 3 nm. However, treatment for more than 6 h lead to more micropores generated and the pore size distribution shifted to the macropore region. This may be attributed to mesoporous merged into macropore after too long treatment time.

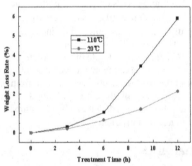

Figure 3. Weight loss ratio of samples

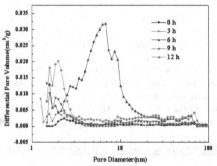

Figure 4. Pore size distribution of samples

Morphology and microstructure

Table II presents the AAS analysis results of the dissolved ions in the acid solutions after treatment at 110☐ for 6 h. As expected, the main dissolved contents are Al^{3+}, Mg^{2+} and Si^{4+} ions because they are the essential constituent elements of cordierite ($2MgO \cdot 2Al_2O_3 \cdot 5SiO_2$). The weight ratio of the main elements Al^{3+}, Mg^{2+} and Si^{4+} in the acid solution is 1:0.26:0.17.

However, according to the initial composition, the ratio should be 1:0.44:1.30. The acid dissolves Al^{3+} and Mg^{2+} ions preferentially. It seems that Si is to be more anti-acid and remain in the acid-treated cordierite structures. Others are impurity ions, including Ca^{2+}, Fe^{3+}, Ti^{4+}, K^+ and Na^+. The relative dissolved amount of Ca^{2+}, K^+ and Na^+ are 16.2%, 19.7% and 25.9% respectively, much higher than that of main elements. It is well known that these impurities, especially K^+ and Na^+, are very harmful to lowering the CTE of cordierite ceramics. Because they are usually prone to form glass phases that have a much higher CTE, the removal of them from the samples explains, at least partially, the reason why acid treatment can greatly decrease the CTE.

Table II. Spectral analysis results of acid treatment at 110℃ for 6 h

Solution	Al^{3+}	Mg^{2+}	Si^{4+}	Ca^{2+}	Fe^{3+}	Ti^{4+}	K^+	Na^+
Concentration(mg/L)	347.6	88.8	59.3	4.8	4.6	3.2	3.4	1.3
Relative dissolved amount(%)	5.8	7.5	0.7	16.2	12.3	5.4	19.7	25.9

After corrosion in the thermal HNO_3 solution, the surface of cordierite solids becomes coarser than untreated. It is shown that cross-section of sample untreated presents densification and less pore in Figure 5(a). As shown in Figure 5(b), a structure of more porous is observed for the sample after treatment for 6 h. According to the results shown in Table II, part of the glass phase in the body is eroded by acid, then the size of pores become larger after being eroded. Additionally, the gaps between grains enlarge which can afford enough free space to expansion under high temperature. However, this is the important reason for the above-mentioned results of quick loss in mass and mechanical strength after acid corrosion. From these data it can be concluded that advantages of acid leaching include the potential for eliminating the washcoat, tailing the pore structure of the cordierite and improving thermal shook resistance.

The XRD spectrum diagrams of samples untreated and treated are shown in Figure 6. All of these XRD patterns match the one reported for cordierite in previous works [1,2,11,12]. These X-ray diffraction patterns show no significant differences between them regarding the number or position of the peaks. However, for the acid corroded cordierite, acid-resistant free amorphous silica is generated on the surface of cordierite solids. This is also testified by a broad underlying structure in the range 2θ of 20-30°, which conforms to the acid leaching results concluded by Elmer [2].

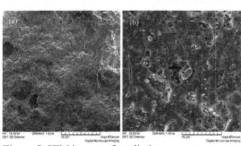

Figure 5. SEM images of cordierites

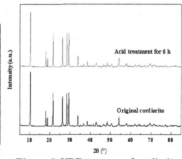

Figure 6. XRD patterns of cordierites

145

CONCLUSIONS

The coefficient of thermal expansion and compression strength of cordierite decrease obviously after acid treatment at 110℃ with extension of acid treatment time. This is mainly because that the destruction reaction of the silicate structure is more drastic at a high temperature. The corrosion content of cordierite is increased and micropores are gradually formed, thereafter leading to an increase of specific surface area. Cordierite shows a bimodal distribution of pore size with two peaks appearing around 1.5nm and 6nm, respectively after acid treatment of 6 h. It is suitable for cordierite matrix acid erosion to afford sufficient mechanical strength and optimal pore structure for DeNOx catalyst matrix at 110℃ for 6 h.

ACKNOWLEDGMENTS

The authors are grateful for the financial support of Chinese National Programs for High Technology Research and Development (No.2010AA065001) and Sci-tech Personnel Service for Enterprise (No. 2009GJF10047).

REFERENCES

1. A.N. Shigapov, G.W. Graham, R.W. McCabe, M. Paputa Peck and H. Kiel Plummer, *Appl. Cata. A.* **182** (1999) 137-146.
2. T.H. Elmer, Selective leaching of extruded cordierite honeycomb structures,*Ceram. Eng. Sci. Proc.* **7** (1986) 40-51.
3. V. Tomasic and F. Jovic, *Appl. Cata. A.* **311** (2006) 112-121.
4. S. Cimino, R. Pirone and G. Russo, *Ind. Eng. Chem. Res.* **40** (2001) 80-85.
5. R.R. Broekhuis, B.M. Budhlall and A.F. Nordquist, *Ind. Eng. Chem. Res.* **43** (2004) 5146-5155.
6. P. Avila, M. Montes and E.E. Miró, *Chemical Engineering Journal*, **109** (2005) 11-36.
7. M. Skoglundh, H. Johanssön, L. Lowendahl, K. Jansson, L. Dahl and B. Hirschauer, *Appl. Catal. B.* **7** (1996) 299-307.
8. C.D. Madhusoodana, R.N. Das, Y. Kameshima, A. Yasumori and K. Okada, *J. Porous Mater.* **8** (2001) 265-271.
9. Q. Liu, Z. Liu and Z. Huang, *Ind. Eng. Chem. Res.* **44** (2005) 3497-3502.
10. J. Bai and L. Guo, *J. Wuhan University of Tech.* **20**(2006) 100-102.
11. H. Galindo, Y. Carvajal and S. L. Suib, *Microporous Mesoporous Mater.* **135**(2010)37-44.
12. Y. Dong, X. Feng, D. Dong, S. Wang, J. Yang, J. Gao, X. Liu and G. Meng, *J. Membrane Sci.* **304** (2007) 65-75.

Mater. Res. Soc. Symp. Proc. Vol. 1449 © 2012 Materials Research Society
DOI: 10.1557/opl.2012.1290

Microwave synthesis of ZrO_2 and Yttria stabilized ZrO_2 particles from aqueous precursor solutions

Kenny Vernieuwe[1], Petra Lommens[1], Freya Van den Broeck[2], José C. Martins[2], Isabel Van Driessche[1], Klaartje De Buysser[1]

[1] SCRiPTS, Department of Inorganic and Physical Chemistry, Ghent University, 218-S3 Krijgslaan, Ghent, B-9000, Belgium

[2] NMR and structure analysis, Department of Organic Chemistry, Ghent University, 218-S4 Krijgslaan, Ghent, B-9000, Belgium

ABSTRACT

Zirconia and Yttrium stabilized zirconia are well-known ceramic materials. Scaling down the dimension of these ceramics can result in a faster sintering process at lower temperatures. Microwave synthesis of nano-structured particles is a very attractive synthesis route because of the short synthesis time and low reaction temperature. This allows a fast screening of the influence of different parameters such as time, temperature and pressure on the final size and crystal phase of the particles. In this study Zr and Zr/Y aqueous precursors are mixed with a variety of complexing agents or surfactants in different ratios. The reason is twofold: (1) we aim for a stable precursor solution which is established by lowering the free ion concentration and (2) we want to see the influence of the complexing agents on the growth of the particles and the formation of crystalline phases. Particle sizes of these particle vary from 40 -200 nm. The crystallinity is confirmed by X-ray diffraction. The stabilization of these particles and possible exchange of the ligands is examined with NMR measurements (1D - proton combined with 2D NOESY) and is compared with TGA-DTA analysis of the isolated particles.

INTRODUCTION

Ceramics are by definition a combination of metallic and non-metallic elements which gain their specific properties by a high temperature treatment. The temperatures necessary for crystallization and formation of the desired phases can be very high resulting in an high energy consumption [1]. In the light of green chemistry and environmental and energy friendly processes a lot of effort has been put in the research towards novel synthesis and preparation routes that overcome this temperature barrier. Well established solid state reactions for formation of ceramic use the shake-and-bake method in order to homogenize the precursor mixture and the increase the inter-granular contacts. Solution chemistry routes avoid those intensive milling and mixing steps and ensure a homogeneity of the precursor to the atomic level [2, 3]. Off course, precipitation has to be prevented at all times to avoid non-stochiometric precursor solutions. Solution chemistry routes allow beside the formation of bulk ceramics, the deposition of ceramic thin films by dip-coating, spin-coating, spray-coating or even ink-jet printing [4-7].

In all the above mentioned systems there is still a need for a high temperature treatment for the formation of the desired crystalline phase. In this work we focus on the synthesis of ceramic nanoparticles from solution precursors. These particles should be as crystalline as possible so that a much lower temperature treatment is necessary to convert the wet phase into a crystalline bulk or thin film material. This will open a new branch of applications such as the deposition of ceramic thin films on polymers. Multiple approaches for the synthesis of nanoparticles from bottom-up have already been published such as: hot-injection, precipitation, peptization, ultrasonic treatment, solvothermal and hydrothermal treatment, microwave assisted solvothermal and hydrothermal treatment and many more [8-13]. In this paper the microwave assisted hydrothermal synthesis of well established ceramics as zirconia (ZrO_2) and yttria-stabilized-zirconia (YSZ) particles is described. Cubic zirconia is only stable above a temperature of 2000°C. Cooling down induced a phase transition to the monoclinic phase and subsequently cracks in the material. Due to these cracks and the different crystal lattice, pure zirconia inhibits the use for many applications. The cubic crystal structure can be retained at room temperature by addition of yttria. YSZ ensures the use for various applications such as thermal barrier coating in engines, SOFC, ceramic knifes, tooth crowns etc. The conventional method to synthesize nano-particulate YSZ is by a hydrothermal process. The long reaction times and high temperatures have practical and economical disadvantages. Microwave assisted hydrothermal synthesis can sometimes lower the reaction temperature but certainly scale-down the reaction times form hours to minutes. In this work the possibility of microwaves is tested for ZrO_2 and YSZ and the influence of different complexing agents of the formation of crystalline phases and particle growth are explored.

EXPERIMENT

ZrO_2 precursor

Zirconium oxynitrate hydrate ($ZrO(NO_3)_2.nH_2O$, Sigma Aldrich) is dissolved in distilled water (0.33 mol/L). The solution is kept in an ultrasonic bath until the salts are completely dissolved. 2.4 w% PEG 1000 (Alfa Aesar) is added to the solution to avoid agglomeration of ZrO_2 on the wall of the vessels. After μ-wave synthesis and purification, citric acid (Carl Roth) and ethanolamine (Sigma Aldrich) are added to stabilize the particles at a neutral pH level.

YSZ precursor

Zirconium oxynitrate hydrate ($ZrO(NO_3)_2.nH_2O$, Sigma Aldrich) and yttrium nitrate ($Y(NO_3)_3.nH_2O$, Alfa Aesar, 99.9%) are dissolved in distilled water. The solution is kept in an ultrasonic bath until the salts are completely dissolved. Ethylenediaminetetraacetic acid (EDTA, Alfa Aesar, 99%) or nitrilotriacetic acid (NTA, Alfa Aesar, 99%) are used as complexing agents. EDTA is used in a 1:1.5 ratio (metal:ligand) whereas NTA in a 1:2 is used. Both solutions (salts and complexing agent) are heated till 60° C before the complexing agent is added drop wise to solution of the salts. Finally, the pH of the mixtures is adjusted by addition of ethanolamine (Sigma Aldrich, ≥99.9%). In some experiments a pH range is explored. The pH is lowered by formic acid (VWR, 100%) or ethanolamine is added for more alkaline solutions. The final concentration equals 0.033 mol/L.

Microwave assisted hydrothermal synthesis

2 mL of the above prepared precursors are put in to capped 10 mL vessels and are treated by microwave (Discovery, CEM) at temperatures between 110 °C – 180 °C for 2 till 10 minutes. The particles are purified by destabilization with an ethyl acetate/ethanol mixture. They are separated by centrifugation and redispersed in an aqueous solution of citric acid and ethanol amine.

Characterization

The particle sizes and particle size distribution are examined by dynamic light scattering (Zetasizer Nano ZS, Malvern). The crystallinity and the crystal phases were determined with X-ray diffraction (Thermo Scientific ARL X'tra diffractometer) followed by Rietveld refinement (TOPAS Academic). Differential and thermogravimetric analysis (DTA, TGA) (STA449F3 Jupiter, Netzsch) are used to examine the decomposition behavior. X-ray fluorescence (XRF) (Rigaku, Nex CG) is used to analyze the amounts of yttrium and zirconium in the YSZ samples. NMR experiments (1D proton, 2D proton and 2D NOESY, Bruker DRX 500 MHz AVANCE) were performed to identify the presence of stabilization groups fixed to the particles.

DISCUSSION

ZrO$_2$ particles

Not all solutions remained stable after microwave assisted hydrothermal treatment. The samples treated at higher temperature and longer times showed precipitation afterwards that could not be redissolved. Figure 1 shows the results of the analysis of the destabilized and dried particles after μ-wave treatment. With increasing temperature the hydrodynamic diameter of the particles shifts from 40 to 60 nm (Figure 1a). The polydispersity is rather broad. X-ray analysis was performed to see whether any crystallinity could be obtained at this low temperatures. As can be seen in Figure 1b, broad reflections are obtained caused by the small crystallite sizes. Monoclinic and to a lesser extend tetragonal zirconia phase could be identified with an average crystallite size of 5 nm. No remarkable changes could be noticed upon variation in temperature and time. The diffraction pattern of the precursor treated at 180 °C for 10 min shows much more crystalline materials but the solution was not stable after μ-wave treatment and precipitation could be seen. No exact percentage of the amorphous phase could been calculated as no internal standard was added to the powders. The particles measured in solution are thus clearly agglomerates of smaller crystallites.

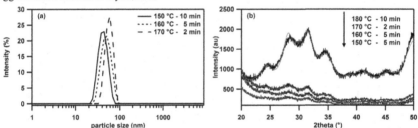

Figure 1: (a) DLS results of the μ-wave treated ZrO$_2$ precursor at various conditions – (b) X-Ray diffraction pattern of the destabilized and dried particles. The red lines indicate the calculated X-ray pattern by Rietveld analysis.

TGA-DTA analysis is performed on the synthesized solutions during the different steps of the purification. Each sample is dried at 60 °C before TGA-DTA analysis. TGA-DTA analysis of a sample after washing with ethanol (Figure 2a) shows a weight loss of approximately 30 % that can be attributed to loss of water, ethanol and PEG that is associated with the particles. After this washing procedure, the particles are redispersed in an aqueous citric acid solution (0.33 mol/L). Figure 2b (thick lines) clearly shows that a lot of organic material is burned off during this procedure even at higher temperatures in comparison with the previous sample. Only 10 % oxide material was remained afterwards. The suspension of particles in the citric acid solution are destabilized, separated by centrifugation and washed several times. TGA-DTA analysis of this sample (Figure 2b thin lines) shows the same decomposition behavior as before washing but now a substantial amount of citric acid is removed. Nevertheless, even after a few washing steps citric acid is remained in the sample or on the surface of the particles. NMR 1D and 2D spectra are used to see whether organic groups are attached to the surface of the particle in order to stabilize the particles and to identify the nature of this organic groups.

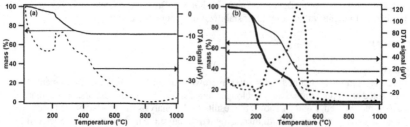

Figure 2: (a) TGA-DTA curve of particles after first washing step. Full line: TGA signal – Dashed line: DTA signal - (b) TGA-DTA curve of particles after first washing step and after addition of citric acid (thick lines) and after an additional washing step (thin lines).

The 1D-NMR spectra show the presence of PEG, citric acid and ethanolamine in the suspension. Nevertheless, this gives no proof for the fact that these groups are actually attached to the surface of the particles. A strong negative nOe cross coupling in the 2D-NOESY spectrum can confirm the presence of the group at the surface of the particles [14]. This can be seen for the ethanolamine groups The non-symmetric signals for citric acid are also caused by interface interactions. The interaction is rather fast most likely due to an exchange with other ethanolamine and citric acid groups in solution. No such behavior could be noticed for the PEG systems.

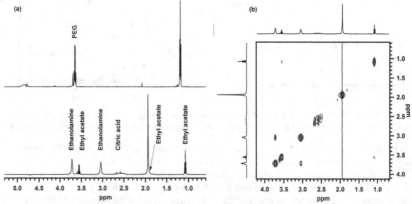

Figure 3: (a) 1D NMR spectra at room temperature, 500 MHz. The upper spectrum corresponds to a suspension directly after μ-wave synthesis. The lower spectrum shows a sample after addition of citric acid and ethanolamine - (b) 2D NOESY NMR spectrum at room temperature, 500 MHz.

YSZ particles

Particle sizes (Figure 4a and 4b) vary from 80-90 nm for the NTA system to 100-190 nm for the EDTA system. The polydispersity is more in the EDTA systems as stronger complexes are formed which limits the particle growth. This variation of the complexing agent limits the temperatures that can be used in the μ-wave set-up. As the complex stability constant for Zr-EDTA complexes is $10^{29.4}$ and Zr-$(NTA)_2$ is $10^{20.9}$ [15], it can be understood that higher temperatures will be necessary to break down the strong complex in order to allow the particles to grow.

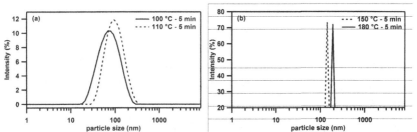

Figure 4: (a) DLS results of the μ-wave treated YSZ NTA - precursor at various conditions – (b) DLS results of the μ-wave treated YSZ EDTA - precursor at various conditions.

The effect of pH on the particle size for the EDTA system is shown in Figure 5a. The particles have the smallest sizes in the pH range 6-7. In this pH interval the EDTA structure mainly exists in its double and triple deprotonated form which facilitates the formation of stronger complexes in comparison with the protonated and single deprotonated species. The question remains if YSZ is formed or mixed ZrO_2 - Y_2O_3 phases. XRF analysis of the isolated particles show that the same ratios Zr: Y can be found as in the as-prepared solution. X-ray analysis of and EDTA stabilized sample cured at 180 °C for 5 min showed that besides the monoclinic (60 w%) and tetragonal (20 w%) phase 20 w% of cubic YSZ could be found. More detailed analysis about the homogeneity and yields need to be performed.

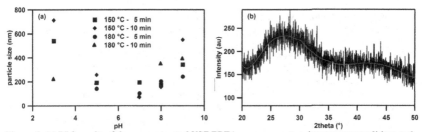

Figure 5: (a) DLS results of the μ-wave treated YSZ EDTA - precursor at various μ-wave conditions and with varied pH of the precursor solution - (b) X-Ray diffraction pattern of a YSZ –EDTA precursor pH 5 and treated at 180 °C for 5 min. The red line indicate the calculated X-ray pattern by Rietveld analysis.

CONCLUSIONS

Microwave assisted hydrothermal treatment has proven to be effective for the synthesis of nanoscaled (40-200nm) ZrO_2 and YSZ particles. The stabilization of zirconia particles in water can be established by addition of citric acid and ethanolamine. These species are present at the surface of the particles but are rapidly exchanged. The addition of complex agents during the microwave assisted hydrothermal synthesis allows a stable precursor solution but variation in the pH strongly affects the particle sizes. Upon deprotonation of the complexing agent, more stable complex are formed with smaller particles as result.

ACKNOWLEDGMENTS

The authors would like to thank prof. dr. Johan Van der Eycken (Ghent University) and prof. dr. S. De Smedt (Ghent University) for use the microwave and dynamic light scattering set-up.

REFERENCES

1. W.D. Callister, in Materials Science and Engineering: An Introduction, (Wiley & Sons, Danvers, 2000), p. 661.
2. C.J. Brinker and G.W. Scherer, in Hydrolysis and Condensation I Nonsilicates, (Academic press, San Diego, California, 1990), p. 20
3. J. Livage, New Journal of Chemistry 25, 1 (2001).
4. J. Feys, P. Vermeir, P. Lommens, S.C. Hopkins, X. Granados, B.A. Glowacki, M. Bäcker, E. Reick, S. Ricart, B. Holzapfel, P. Van Der Voort and I. Van Driessche, J. Mater. Chem. 22, 3717 (2012).
5. M. Arin, P. Lommens, N. Avci, S.C. Hopkins, K. De Buysser, I.M. Arabatzis, D. Poelman and I. Van Driessche, J. Eur. Ceram. Soc. 31, 1067 (2011).
6. M. Arin, P. Lommens, S.C. Hopkins, G. Pollefeyt, J. Van der Eycken, S. Ricart, X. Granados, B.A. Glowacki and I. Van Driessche, Nanotechnology (2012).
7. V. Cloet, J. Feys, R. Huhne, S. Hoste and I. Van Driessche, J. Solid State Chem. 182, 37 (2009).
8. T. Okubo and H. Nagamoto, J. Mater. Sci. 30, 749 (1995).
9. H. Armendariz, M.A. Cortes, I. Hernandez, J. Navarrete and A. Vazquez, J. Mater. Chem. 13, 143 (2003).
10. E. Gressel-Michel, D. Chaumont and D. Stuerga, J. Colloid Interface Sci. 285, 674 (2005).
11. M.L. Moreira, G.P. Mambrini, D.P. Volanti, E.R. Leite, M.O. Orlandi, P.S. Pizani, V.R. Mastelaro, C.O. Paiva-Santos, E. Longo and J.A. Varela, Chem. Mat. 20, 5381 (2008).
12. E.T. Thostenson and T.W. Chou, Compos. Pt. A-Appl. Sci. Manuf. 30, 1055 (1999).
13. M. Zawadzki, Journal of Alloys and Compounds 454, 347 (2008).
14. A. Hassinen, I. Moreels, C.D. Donega, J.C. Martins and Z. Hens, J. Phys. Chem. Lett. 1, 2577
15. A.E. Martell and R.M. Smith, in Critical Stability Constants - Volume 3: Other Organic Ligands edited by Editor (Plenum Press, New York, 1977), p. Pages.

Mater. Res. Soc. Symp. Proc. Vol. 1449 © 2012 Materials Research Society
DOI: 10.1557/opl.2012.1223

Synthesis of water dispersed Fe₃O₄@ZnO Composite Nanoparticles by the Polyol Method

Yesusa Collantes[1], Oscar Perales-Perez[2], Oswald N. C. Uwakweh[2] and Maxime J.-F.Guinel[3]
[1]Department of Physics, University of Puerto Rico, Mayaguez, PR 00680 USA
[2]Department of Engineering Science and Materials, University of Puerto Rico, Mayaguez, PR 00680 USA
[3]Department of Physics, University of Puerto Rico, San Juan, PR 00936-8377 USA

ABSTRACT

Water-soluble Fe_3O_4@ZnO composite nanoparticles (NPs) were synthesized using a polyol route. The effects of the addition of the ZnO phase were evaluated by varying the Zn/Fe molar ratio in the 0.25-1.00 range as a function of the reaction time. X-ray diffractometry confirmed the formation of the magnetite and ZnO phases and suggested the possible formation of a composite structure. Also, using this method, pure magnetite and ZnO NPs were synthesized. The average crystallite sizes were estimated to 6.3 ± 0.3 nm and 8.6 ± 0.6 nm for magnetite and ZnO NPs, respectively. The samples were examined using transmission electron microscopy. Fourier transform infrared spectra indicated the presence of adsorbed species onto the solids surface, which may explain the good stability of the materials in water. Photoluminescence measurements at room temperature for pure ZnO nanoparticles exhibited the characteristic excitonic emission around 395 nm. Vibrating Sample Magnetometer measurements at room temperature evidenced the superparamagnetic behavior of magnetite nanocrystals, with a saturation magnetization of 60emu/g. The maximum magnetization ranged from 28 to 54emu/g for the composite NPs. Mössbauer spectroscopy measurements at room temperature showed evidence of evolving Fe-sites associated to superparamagnetic particles, as reflected on the coexistence of prominent doublet peaks and very weak sextet peaks.

I.-INTRODUCTION

Composite NPs have attracted much attention due to their unique physico-chemical properties and potential applications. These heterostructured systems present multifunctional properties arising from the synergy between co-synthesized materials [1]. The use of these NPs for Photo-dynamic Therapy (PDT) relies on the fact that semiconductor nano-materials could generate singlet oxygen and becomes a new generation of photosensitizers (PS) [2]. ZnO is an excellent PS candidate due to its nontoxicity and ability to biodegrade [3,4]. Moreover, it displays high thermal and chemical stabilities [5]. Furthermore, ZnO is a high quality semiconductor material with a band gap of 3.4 eV, is transparent in the UV region and has a large excitation binding energy at room temperature (60 meV) [6]. ZnO has the hexagonal wurtzite structure with lattice parameters $a = 3.29$Å, and $c = 5.24$ Å [6]. In turn, superparamagnetic magnetite combines high magnetic susceptibility and saturation magnetization with low remanence and low coercivity, finding multiple applications in bioseparation, targeted drug delivery, immunoassays and biosensors [7]. Therefore, the development of a composite Fe_3O_4@ZnO structure could provide new avenues as a biocompatible and bi-functional platform with suitable optical properties to be applied as a magnetic PS in PDT or in magnetic targeted delivery. The present work is focused on the synthesis and characterization of Fe_3O_4@ZnO composite NPs synthesized through a simple two-stage approach using a polyol medium. The polyol medium acted as the solvent for the formation of NPs while enhancing their water solubility [8, 9]. The corresponding structural and optical properties were discussed as a function of the Zn/Fe molar ratio in the 0.25-1.00 range and the reaction time.

II.-EXPERIMENTAL DETAILS

A. Materials

Iron (III) acetylacetonate [Fe(acac)$_3$, 99%] and zinc acetate dehydrate [Zn(OOCCH$_3$)$_2$.2H$_2$O, 99%] were used as precursors salts. Triethylene glycol (TREG, 99%) was the solvent. The total ions concentration was fixed at 0.08 mol/L in all experiments.

B. Synthesis of composite nanoparticles

The Fe$_3$O$_4$@ZnO composite NPs were synthesized by the sequential precipitation of ZnO onto the surface of magnetite seeds. The synthesis of the magnetite NPs took place by heating a Fe(acac)$_3$ solution in TREG at 280°C and kept under reflux and vigorous stirring for 30 min. The suspension was then cooled down to room temperature, yielding a black suspension containing the Fe$_3$O$_4$ NPs. In the second stage, appropriate amounts of Zn(OOCCH$_3$)$_2$.2H$_2$O were added to the magnetite suspension in order to adjust the Zn/Fe molar ratio in the 0.25-1.00 range and control the amount of ZnO. The resulting mixture was slowly heated to 280 °C and kept under reflux for 30 min. The colloidal suspension of Fe$_3$O$_4$@ZnO composite obtained was centrifuged at 8,000 rpm for 20 minutes in order to remove as much as possible the excess of solvent, washed with ethanol three times and re-dispersed in water. A similar procedure was followed to investigate the formation of ZnO in absence of magnetite seeds. Composite NPs were recovered by centrifugation, dried at 60 °C and characterized.

C. Characterization

Powder x-ray diffractograms were recorded using a Siemens D500 operating with a copper target (Cu-K$_\alpha$ radiation). Average crystallites sizes were calculated using the main diffraction peaks and according to Scherrer's equation [10]. The presence of absorbed surface species was determined using Fourier Transform Infrared spectra, recorded using a Shimadzu IR-Afinity spectrometer. The materials were examined using a JEOL JEM-2200FS high resolution transmission electron microscope, also equipped with an x-ray energy dispersive spectrometer. The optical properties were measured using a Fluoromax2 Photo-spectrometer. The excitation wavelength for Photoluminescence measurements was set at 350 nm. Room-temperature hysteresis loops were recorded using a Lake Shore 7410 vibrational sample magnetometer. Mössbauer measurements were carried out using a Webres spectrometer (Ritverc, GmbH) operating in the transmission mode, operated with a cobalt source in a rhodium matrix.

III.-RESULTS AND DISCUSSION

A. X-ray diffraction (XRD)

The x-ray diffractograms recorded from the Fe$_3$O$_4$ NPs, the ZnO NPs and the Fe$_3$O$_4$@ZnO composite NPs are shown in Fig. 1 (a) and (b) for different Zn/Fe molar ratios and reaction times, respectively. All peaks were indexed to Fe$_3$O$_4$ magnetite and ZnO wurtzite. Their coexistence suggests the formation of a composite nanostructure [11]. The average crystallite sizes were calculated to be 6.3 ± 0.3 nm for magnetite and 8.6 ± 0.6 nm for ZnO.

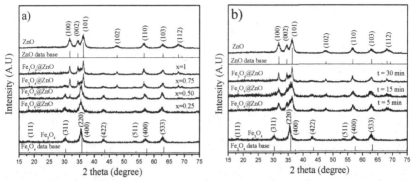

Fig. 1. XRD patterns recorded for materials synthesized using different x=Zn/Fe molar ratios for 30 min. (a), and reaction times with a fixed Zn/Fe molar ratio equal to unity, (b).

B. Fourier Transform Infrared spectroscopy (FTIR)

FTIR spectra recorded from the pure TREG and the materials obtained using a Zn/Fe molar ratio =1 and a reaction time of 30 minutes are shown in Fig. 2. The presence of organic moieties on the surface of the materials associated to the functional groups of polyol was detected. This is confirmed with the presence of the bands at 2,925-2,809 cm^{-1} (C-H stretching, ester group) and 1,116 – 1,050 cm^{-1} (C–O stretching, alcohol group) [8]. The broad band centered at 3,351 cm^{-1} corresponds to the O–H stretching vibration in surface hydroxyl groups [8]. The intense band at 585 cm^{-1} can be attributed to the Zn-O and Fe-O bonds in magnetite and ZnO [12]. The band at 1,630 cm^{-1} is assigned to the presence of residual water (H$_2$O) in the sample [13]. These surface groups provide stability to the NPs when suspended in water because of the 'steric' repulsion between the NPs and a positively charged surface [14].

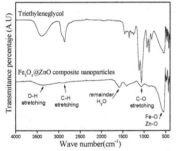

Fig. 2. FTIR spectra of the pure TREG and the Fe$_3$O$_4$@ZnO composite NPs synthesized using a Zn/Fe molar ratio= 1 and 30 minutes of reaction.

C. Transmission electron microscopy (TEM)

The samples were examined in the TEM and also in the scanning mode (STEM). Fig. 3 is a montage of 4 representative images showing a ZnO NP (a), and an overview of the Fe$_3$O$_4$@ZnO composite NPs synthesized at Zn/Fe=0.25 and 30 minutes (b, the inset is a FFT). Two HRTEM images of the composite NPs are shown in (c, d). Despite the agglomeration, there was presence of monodispersed nanoparticles with a diameter below 10nm; however, the presence of residual solvent in the samples prevented high magnification STEM images to be recorded. It was not possible to record x-ray energy dispersive spectrometry (XEDS) measurements or line scans from single composite NPs. These analyses could have determined unambiguously the core-shell or composite nature of the composite, explaining the choice for the 'composite' appellation. Probing several large agglomerates in the same sample yielded the mean composition: 84 at. % Fe and 16 at. % Zn. Similar results were obtained from agglomerates of few particles. A typical XEDS is shown in (e).

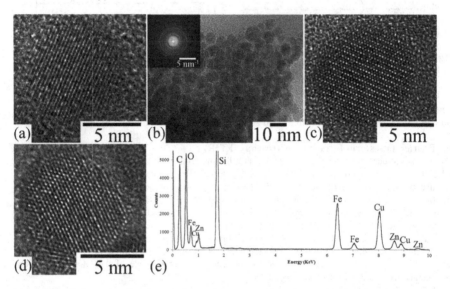

Fig. 3. TEM images showing (a) a ZnO NP; (b) an overview of the Fe_3O_4@ZnO composite NPs where the inset is a FFT; (c, d) two composite NPs. A XEDS collected from a group of NPs is shown in (e). The mean composition was determined to be 84 at. % Fe and 16 at. % Zn.

D. Photoluminescence (PL) Measurements

Fig. 4 displays the PL spectra recorded from the ZnO and the composite NPs. Pure ZnO powders exhibited a strong ultraviolet (UV) emission peak centered at 395 nm, which is due to an exciton related recombination near the band-edge of ZnO [5]. In both set of spectra, PL peaks for Fe_3O_4@ZnO nanocomposite were blue-shifted with respect to the emission peak of pure ZnO. This shift could be attributed to the charge transfer at the Fe_3O_4@ZnO interface [15]. Further experiments are in progress to examine this fact in detail. The blue shift is less prominent when the x=Zn/Fe molar ratio is increased, the same effect is observed for prolonged reaction times of ZnO formation, this could result from increasing amounts of ZnO [15] which is verified by XRD analysis, where the ZnO peaks are more intense for larger x=Zn/Fe and prolonged reaction times.

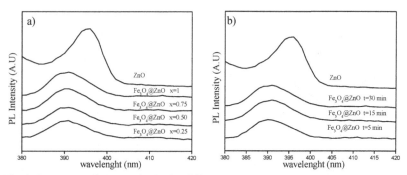

Fig. 4. PL spectra of NPs synthesized at different x= Fe/Zn molar ratios for a reaction time of 30 minutes, (a), and different reaction times for x=1, (b).

E. Magnetic Measurements

Room-temperature hysteresis loops were recorded from pure magnetite and $Fe_3O_4@ZnO$ composites synthesized using different x=Zn/Fe molar ratios and 30 minutes of reaction are shown in Fig. 5 (a). The loops recorded with various reaction times and a fixed molar ratio x=1 are shown in (b). The saturation magnetization values were in the 28-60 emu/g range. The lower values were observed in samples synthesized with higher Zn/Fe molar ratios or prolonged reaction times. This drop in magnetization for the composite NPs is attributed to the enhanced presence of diamagnetic ZnO co-existing with magnetite [16]. Moreover, M-H measurements revealed the superparamagnetic nature of the materials due to the nano-sized magnetite seeds [7].

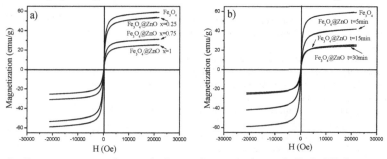

Fig. 5. Room-temperature hysteresis loops for magnetite and $Fe_3O_4@ZnO$ composites synthesized using different x=Zn/Fe molar ratios for 30 minutes of reaction, (a), and for various reaction times at x=1, (b).

F. Mössbauer Spectroscopy

Fig. 6 shows the room-temperature Mössbauer profiles corresponding to magnetite and $Fe_3O_4@ZnO$ composite synthesized using x=Zn/Fe molar ratios equal to unity and 30 minutes of reaction. The absence of hyperfine splitting and the exclusive presence of the central doublet suggest the presence of Fe-sites associated with a superparamagnetic ordering. The probable formation of a double oxide structure of the type $ZnFe_2O_4$ at the magnetite/ZnO interface can also be suggested because of the slight but noticeable changes in the spectral peaks line positions and intensities, coupled with the broadening around the central portions of Fe_3O_4@ZnO spectra in comparison to magnetite alone.

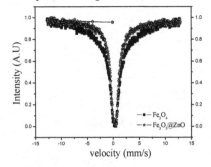

Fig.6. Room-temperature Mössbauer profiles corresponding to magnetite and Fe_3O_4@ZnO composite synthesized at x=Zn/Fe molar ratio =1 and 30 minutes of reaction.

CONCLUSIONS

The probable formation of Fe_3O_4@ZnO composite NPs was investigated using XRD, spectroscopy, TEM, VSM and Mössbauer measurements. Adjusting the Zn/Fe molar ratio or the reaction time increased the amount of ZnO that could explain the observed changes in the structural, optical and magnetic properties of the composite material. Room temperature magnetization and Mössbauer measurements revealed the superparamagnetic nature of the composite NPs.

Acknowledgments

This material is based upon work supported by the National Science Foundation under Grant No. HRD 0833112 (CREST program at UPRM).

References

[1] Guannan Wang and Xingguang Su, Anal. **136**, 1783, (2011).
[2] P. Juzenas, et al., Adv. Drug. Deliv. Rev. **60**, 1600, (2008).
[3] Z. L. Wang, et al., Adv. Funct. Mater. **14**, 943, (2004).
[4] J. Zhou, et al., Adv. Mater. **18**, 2434, (2006).
[5] E. Çetinörgü and S. Goldsmith, J. Phys. D: Appl. Phys. **40**, 5220 (2007).
[6] Ü. Özgür, et al., J. Appl. Phys. **98**, 5220, (2005).
[7] T.K. Indira and P.K. Lakshmi, Int. J. Pharm. Sci. Nanotech. **3**, 1035, (2010).
[8] Wei Cai, et al., J. Colloid Interface Sci. **305**,366, (2007)
[9] E. V. Panfilova, et al., Colloid J. 74, 99, (2012)
[10] B. D Cullity, in *Elements of X-ray Diffractions*, edited by Morris Cohen (Addison Wesley, MA, 1972), p. 102.
[11] Jiaqi Wan, et al., Mater. Chem. Phys. **114**, 30, (2009).
[12] R. M. Cornell and U. Schwertmann, in *The Iron Oxides*, (Wiley, Weinheim, 2003), p.141.
[13] T. Rajh, et al., J. Phys. Chem. B **106**, 10543, (2002).
[14] Junhua Zhao, et al., Am. Ceram. Soc. Bull. **94**, 725, (2011).
[15] Peng Zou, et al., J. Nanosci. Nanotechnol. **10**,1992, (2010).
[16] R. Y. Hong, et al., Mater. Res. Bull. **43**, 2457, (2008).

Mater. Res. Soc. Symp. Proc. Vol. 1449 © 2012 Materials Research Society
DOI: 10.1557/opl.2012.871

Binding Mechanisms of As(III) on Activated Carbon/Titanium Dioxide Nanocomposites: A potential method for arsenic removal from water

Z. Özlem Kocabaş, Burcu Açıksöz and Yuda Yürüm

Faculty of Natural Science and Engineering, Sabanci University, Orhanlı 34956 Tuzla, Istanbul/Turkey

ABSTRACT

Novel nanocomposite materials where titanium dioxide nanoparticles were inserted into the walls of a macroporous activated carbon were produced and their efficiency for the removal of As(III) from water was compared with pure activated carbon and titanium dioxide nanoparticles. The nanocomposites were synthesized with different molar ratios by using sol-gel method and were characterized by x-ray diffraction (XRD) and scanning electron microscopy (SEM). The nanocomposite system showed excellent capability for the removal of As(III) ions from water by considering feasibility, efficiency and cost. The maximum As(III) removal percentages were ~4.7% at pH 8 for activated carbon, ~38 % for titanium dioxide at pH 6, and ~98 % at pH 7 for activated carbon/titanium dioxide (AC/TiO$_2$) nanocomposite, respectively. According to kinetic sorption data, the higher regression coefficients (R^2) were obtained after the application of pseudo-second order to the experimental adsorption data for all adsorbent materials. The equilibrium data were modeled with the help of Langmuir and Freundlich equations. Overall, the data are well fitted with both the models, with a slight advantage for Langmuir model. The maximum arsenic uptake (q_{max}) value computed from slope of the linearized Langmuir plot was 26.62 mg/g for the adsorption of As(III) onto AC/TiO$_2$ nanocomposite.

INTRODUCTION

Arsenic is one of the well-known toxic contaminants found in the environment [1]. The source of arsenic contamination in water can be classified as natural and anthropogenic activities including geochemical and natural weathering reaction, volcanic emissions, mining, agricultural activities and industrial wastes [2, 3]. The excessively high arsenic concentration especially in drinking water is a challenging water pollution problem for many countries including the USA, China, Chile, Bangladesh, Taiwan, Mexico, Argentina, Poland, Canada, Hungary, Japan, India and Turkey [4]. Arsenic is severely harmful to the human health and long term exposure to arsenic can lead to cancer of the lungs, skin, kidney and liver [5]. Nowadays, removal of arsenic by adsorption is acquired importance due to its technical simplicity and applicability in rural areas, where people are more subjected to polluted drinking water with arsenic [6]. Titanium dioxide is the widely used photocatalyst due to its strong oxidizing power and favorable band gap energy. It is capable of converting As(III) to less toxic and more adsorbable As(V) species under UV irradiation. However, titanium dioxide has difficulties in continuously converting arsenic species under visible light. To overcome this difficulty, in this paper the titanium dioxide nanoparticles are combined with various amount of activated carbon. Moreover, the influence of contact time and pH on As(III) adsorption by using pure titanium dioxide and the nanocomposites are analyzed.

EXPERIMENTAL

Titanium dioxide nanoparticles were synthesized by sol-gel method using titanium tetraisopropoxide (TTIP) ($C_{12}H_{28}O_4Ti$, Aldrich, 97%) as a metal organic precursor. TTIP (5 ml) was added to mixture of distilled water and 2-propanol (C_3H_8O, Merck, 99%). The gel preparation process was started when the pH of solution was adjusted to ~1 by the addition of 1 M HNO_3 under continuous stirring at 80˚C. After the color of solution became transparent, different amount of activated carbon was added to the solution. The obtained solids were dried for several hours at 100˚C and annealed at 400˚C for 2 h.

A standard As(III) solution containing (50 mg/l) was prepared by dissolving 0.066 g As(III) (As_2O_3, Sigma-Aldrich, 99.9%) in 10 mL 1% (w/v) NaOH and making up to 1 L with deionized water. Batch adsorption experiments were conducted in conical flasks shaken in an incubator shaker at 150 rpm mixing rate for 24 hours at 25˚C. All batch experiments were performed using adsorbent material of 0.5 g/l and an initial arsenic concentration of 5 mg/l, except for the isotherm experiments in which the initial arsenic concentration was varied from 0.375 to 20 mg/l. The solution pH was adjusted by addition of 1M HCl or 1M NaOH solutions. At the end of each experiment the solution was separated from the solid adsorbent by using 0.45 μm PVDF membrane filter. Arsenic concentration of the solutions was measured with a Varian, Vista-Pro CCD simultaneously inductively coupled plasma ICP-OES spectrophotometer. Samples before and after adsorption experiments were analyzed to obtain residual arsenic concentration.

DISCUSSION

Characterization of activated carbon/titanium dioxide nanocomposites

SEM images of synthesized titanium dioxide and AC/TiO_2 nanocomposite are indicated in Figure 1 (a) and (b). XRD spectra of titanium dioxide nanoparticles and AC/TiO_2 nanocomposites (Figure 1(c)) are demonstrated the presence of the main (101) peak of anatase at 2θ values of 25.4˚ [7].

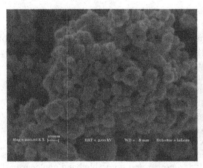

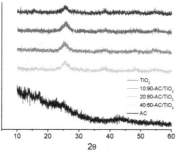

Figure 1: a) SEM micrograph of titanium dioxide nanoparticles **(b)** SEM micrograph of activated carbon/titanium dioxide nanocomposite with composition of 40:60 w/w **(c)** X-ray diffractogram of the titanium dioxide nanoparticles, activated carbon and AC/TiO₂ nanocomposites

Effect of pH

Figure 2 shows the effect of pH on the adsorption efficiency of As(III) on TiO_2 nanoparticles, activated carbon, and the nanocomposites of AC/TiO₂. Combining activated carbon with TiO_2 can increase the separation of the electron-hole charges under visible light and more amount of As(III) species are converted to As(V) species which are more easily removed from water. The maximum As(III) removal percentages are found ~98 % for the AC/TiO₂ nanocomposite with composition of 10:90 w/w at pH 7, respectively. The increase in the activated carbon ratio in the nanocomposite leads to decrease in adsorption performance due to lack of sufficient photocatalytic active sites.

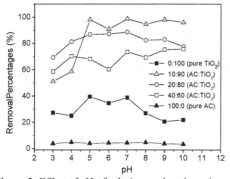

Figure 2: Effect of pH of solution on the adsorption of As(III) with using TiO_2 nanoparticles, activated carbon, and the nanocomposites of AC/TiO₂ with composition of 10:90, 20:80, and 40:60 w/w.

The ionic character of the arsenic species varies with pH as the predominant As(III) species available are H_3AsO_3 and $H_2AsO_3^-$ in the pH range of 4.0-9.5 [8]. The AC/TiO$_2$ nanocomposites are most effectively adsorbed As(III) at pH 7. One likely reason for the observed result is that at pH 7 H_3AsO_4 is mostly found in the aqueous solution and those neutral As(III) species forms complex with surface groups especially at pH in the range of 6.15-8.0 [2].

Sorption kinetics

The effect of time on the removal of As(III) ions from water are investigated for TiO$_2$ nanoparticles, activated carbon, and the AC/TiO$_2$ nanocomposites by taking subsamples at different time intervals. The experiments have been performed with 0.5 g of adsorbent material and 5 mg/l As(III) solution for 5, 10, 20, 40, 60, 120, 240, 480, 720 and 1440 min reaction times. The results indicated in Figure 3 reveal that the adsorption of arsenic species is greatly dependent on contact time. The ~97% of As(III) is removed within 5 min of contact time using AC/TiO$_2$ nanocomposite with composition of 10:90 w/w at pH 7. However, only ~4.8 % of As(III) is removed by using pure activated carbon at 1440 min.

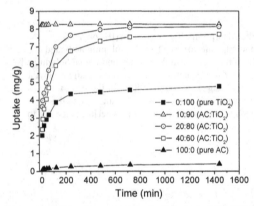

Figure 3: Effect of contact time on the adsorption of As(III) with using TiO$_2$ nanoparticles, activated carbon, and the nanocomposites of AC/TiO$_2$ with composition of 10:90, 20:80, and 40:60 w/w.

The kinetic data obtained are analyzed by employing the pseudo-first order *(1)* and the pseudo-second-order *(2)* equations [9]

$$log(q_1 - q_t) = log(q_1) - (k_{ad}.t) / 2.303 \qquad (1)$$
$$t/q_t = (1/k_2q_2^2) + t/q_2 \qquad (2)$$

Where, q_t is the amount of arsenic adsorbed (mg/g) at time t, q_1 *and* q_2 are the maximum adsorption capacity (mg/g) for the pseudo first-order adsorption and pseudo second order adsorption, k_{ad} is the pseudo-first-order rate constant for the arsenic adsorption process (min^{-1}), k_2 is the pseudo second order rate constant (g/mg min). The results are listed in Table 1.

Table 1. Kinetic parameters for adsorption of As(III) on TiO_2 nanoparticles, activated carbon, and the AC/TiO_2 nanocomposites

Adsorbent	q_1 (mg/g)	k_{ad} (min^{-1})	R^2	q_2 (mg/g)	k_2 (g/mg min)	R^2
0:100 TiO_2	3.705	4.760	0.816	4.803	0.009	0.999
10:90 (AC-TiO_2)	8.254	0.0165	0.285	8.255	1.469	0.999
20:80 (AC-TiO_2)	6.875	5.416	0.739	8.035	0.007	0.999
40:60 (AC-TiO_2)	6.185	9.238	0.872	7.830	0.004	0.999

Adsorption isotherms

The adsorption of As(III) onto TiO_2 nanoparticles, activated carbon, and the AC/TiO_2 nanocomposites over the initial arsenic concentration ranging from 0.375 to 20 mg/l have been performed with using 0.5 g/l adsorbent material in the batch mode. The relationship between the adsorption capability q_e and equilibrium concentration C_e of As(III) onto adsorbent materials is shown in Figure 4.

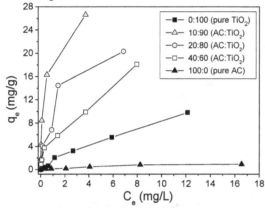

Figure 4: Adsorption isotherm fitted for As(III) adsorption onto TiO_2 nanoparticles, activated carbon, and the AC/TiO_2 nanocomposites

The theoretical adsorption capacity of an adsorbent can be obtained through the adsorption isotherm. In this study, the Langmuir and Freundlich isotherm models were investigated to evaluate adsorption patterns of As(III) on adsorbent materials with respect to its concentration of equilibrium in solution. The Langmuir model assumes monolayer adsorption onto homogenous surface with a fixed number of energetically identical sites [10], while the Freundlich model is derived from the multilayer adsorption and the adsorption occurs first the most energetically favourable sites [11]. The linearized forms of the Langmuir and Freundlich isotherms are

$$C_e / q_e = 1 / (q_{max} L) + C_e / q_{max} \qquad (3)$$
$$\ln q_e = \ln K_f + 1/n \ln C_e \qquad (4)$$

where q_e is the amount adsorbed on solid (mg/g), C_e is the equilibrium solution concentration (mg/l), q_{max} is adsorption capacity (mg/g), L is a constant related to enthalpy of sorption which should vary with temperature (l/mg), K_f, (mg/g) is related to the adsorption capacity of the adsorbent and 1/n is a constant known as the heterogeneity factor is related to surface heterogeneity.

Table II: Calculated isotherm parameters for As(III) adsorption on TiO_2 nanoparticles and the AC/TiO_2 nanocomposites at 25 °C

Adsorbent	q_{max} (mg/g)	L (1/mg)	R^2	K_f (mg/g)	n	R^2
0:100 TiO_2	15.31	0.119	0.412	1.501	1.335	0.938
10:90 (AC-TiO_2)	27.64	6.030	0.992	32.55	0.927	0.595
20:80 (AC-TiO_2)	22.35	1.200	0.927	7.319	0.954	0.722
40:60 (AC-TiO_2)	18.33	0.787	0.850	4.112	1.075	0.737

CONCLUSIONS

The titanium dioxide nanoparticles are combined with different amounts of activated carbon in order to convert As(III) species to less toxic and more adsorbable As(V) species under visible light. By using a sol-gel method, activated carbon/titanium dioxide nanocomposites were successfully synthesized. Adsorption experiments were performed for all adsorbent materials to obtain optimum pH and contact time. The arsenic concentration in the treated water can be reduced below the maximum concentration level requirement (10 μg/l) by using 0.5 g/l of the AC/TiO_2 nanocomposite with composition of 10:90 w/w at pH 7. The increase in the activated carbon ratio in the nanocomposite cause to decrease in removal efficiency of As(III) due to lack of sufficient photocatalytic active sites.

REFERENCES
1. S.H. Lin and R.S. Juang, *J. Hazard. Mater.* **92**, 3 (2002).
2. D. Mohan, C.U. Pittman, *J Hazard Mater.* **142**, (2007).
3. S. Shevade, R.G. Ford, *Water Res.* **38**, (2004).
4. M. Berg, H.C. Tran, T.C. Nguyen, H.V. Pham, R. Schertenleib, W. Giger, *Environ Sci Technol.* **35**, (2001).
5. H.Y. Chiou, Y.M. Hsueh, K.F. Liaw, S.F. Horng, M.H. Chiang, Y.S. Pa, J.S.N. Lin, C.H. Huang, C.J. Chen, *Cancer Res.* **55**, (1995).
6. M. Habuda-Stanic, B. Kalajdzic, M. Kules, N. Velic, *Desalination* **229** (2008).
7. I.R. Bellobono, A. Carrara, B. Barni, and A. Gazzotti, *J. Photoc. Photobio. A.* **84**, 1 (1994).
8. D. Qu, J. Wang, D.Y. Hou, Z.K. Luan, B. Fan, C.W. Zhao, *J Hazard Mater.* **163** (2009).
9. N. Kannan and M.M. Sundaram, *Dyes Pigm.* **51**, 1 (2001).
10. I. Langmuir, *J. Am. Chem. Soc.* **40**, 9 (1918).
11. S.M. Hasany, M.M. Saeed, and M. Ahmed, *J. Radioanal Nucl. Ch.* **252**, 477 (2002).

Mater. Res. Soc. Symp. Proc. Vol. 1449 © 2012 Materials Research Society
DOI: 10.1557/opl.2012.1367

Systematic investigation of the aqueous processing of CdSe quantum dots and CuS nanoparticles for potential bio-medical applications.

Raquel Feliciano Crespo[1], Oscar Perales-Perez[2], Sonia J. Bailon-Ruiz[2] and Maxime J-F Guinel[3]

1 Department of Chemistry, University of Puerto Rico, Mayaguez, PR.
2 Department of Engineering Science & Materials, University of Puerto Rico, Mayaguez, PR.
3 Department of Physics, College of Natural Sciences, University of Puerto Rico, San Juan, PR.

ABSTRACT

Semiconductor quantum dots are considered very promising candidates for bio-imaging and diagnosis applications because of their tunable optical properties and good optical stability in aqueous phase. Any practical application of these materials will rely on the viability of their simple and direct synthesis in aqueous phase with no need for toxic and unstable organic media. The optical properties of CdSe quantum dots and CuS nanoparticles are desirable in bio-imaging and cell sorting applications because of their tunable photoluminescence at the visible range. The present work addresses the synthesis of CdSe quantum dots and CuS nanoparticles via an optimized, simple and scalable aqueous processing route at low temperatures. The tunability of the optical properties was achieved by a suitable control of the citrate/Cd mole ratio, temperature of synthesis (20-90°C) and reaction time (0-1 hour). In the case of CuS, the strong plasmonic absorption offers the opportunity to investigate this material as a photothermal coupling agent for photothermal therapy. The intensity of the plasmonic absorption was enhanced by selecting an appropriate sulfide precursor (Na_2S, Thioglycolic acid), temperature of synthesis (90-120°C) and reaction time. Nanocrystals were characterized by x-ray diffraction, UV-VIS, photoluminescence (PL) spectroscopy techniques and electron microscopy. The effects of the synthesis conditions on the crystal size and the corresponding functional properties of synthesized quantum dots are presented and discussed.

I. INTRODUCTION

Nano particles (NP) and quantum dots (QDs) exhibit unique optical properties that make them promising candidates for nanomedicine applications, such as bioimaging, photodynamic therapy (PDT) and photo-thermal ablation (PTA) [1,2]. These novel nanotechnologies require stable and high quality water-soluble quantum dots [3]. However, the direct synthesis of water-soluble QDs, e.g. CdSe and CuS, exhibiting high stability in biological environments is a very challenging task. To overcome this limitation, the uses of biocompatible capping agents to stabilize QDs are suggested. Citrate ligand species have been used to control crystal growth and promote the quantum yield in semiconductor QDs [4,5]. Although the synthesis of QDs using organometallic complexes routes have been reported elsewhere, the reports on the direct synthesis in aqueous phase of water-stable nanocrystals, e.g. CdSe, are scarce or preliminary. Several studies attributed the toxicity of CdSe QDs to surface oxidation of the nanocrystals or removal of the capping species. However, the mechanism of how the QDs induce toxicity at a cellular level remains unclear. CdSe QDs also have a great potential to biological application as luminescent biological labels and advanced sensors for ex-vivo studies [6,7]. On the other hand, photo-thermal ablation has become an attractive alternative therapy for cancer treatment [7,8]. Among different candidate materials, stable and inexpensive CuS nanocrystals are being considered as photo-thermal agents due to their capability to convert optical energy to thermal energy. CuS exhibits a strong absorption band between 900-1,000 nm, which corresponds to the near-infrared region in the visible spectrum [9]. The lack of systematic research efforts on the size-controlled synthesis of CuS nanocrystals justifies the search of simple and scalable aqueous processing routes to synthesize this material. On this basis, the present work addresses the development of aqueous-based direct synthesis methods for CdSe QDs and CuS NPs under size-controlled conditions.

II. EXPERIMENTAL PROCEDURE

A. Materials

Cadmium acetate [$Cd(CH_3CO_2)_2.2H_2O$, purity 98%] and tri-sodium citrate [$Na_3C_6H_5O_7.2\ H_2O$, purity 98%] from Fisher Scientific, Copper (II) chloride [$CuCl_2\ 2H_2O$, purity 99%], thioglycolic acid [TGA, purity 98%], from Sigma Aldrich, and sodium sulfide [$Na_2S.9H_2O$, crystalline certified ACS] purchased from Fisher Scientific were used as the precursor salts. N,N-dimethylformamide [$HCON(CH_3)_2.$ DMF, purity 99%] was supplied by Across Organic whereas 2-propanol [C_3H_8O] and selenium metal powder [99.99% trace metal basis] were purchased from Fisher Scientific. Deionized water (18MΩ) was obtained using Barnstead systems.

B.   Synthesis of Citrate-CdSe QDs.

The systematic study of the aqueous synthesis of CdSe QDs was based on the protocol described by *P.P. Ingole et al.* [1]. Here, we explored a different reaction sequence and synthesis parameters. DMF and tri-sodium citrate were added to a 0.25mM cadmium acetate aqueous solution and mixed together for one hour at room temperature. The citrate concentration was varied from 0.25mM to 1.3mM, with citrate/Cd mole ratios between 1 and 5. After the precursor salts were dissolved they were contacted with 0.1mM of selenide solution, produced by reductive leaching of elemental selenium powder. The synthesis was carried out at room temperature and at 60°C using a laboratory microwave furnace Mars Xtraction, CEM with power of 400W. The Se/Cd mole ratio was kept constant at 1.0 in all experiments. The reaction time was varied from 0 to 30min in order to optimize the CdSe optical properties. Solids were coagulated with 2-propanol, centrifuged, dried at room temperature and characterized.

C.   Synthesis of CuS NPs

Copper sulfide NPs were synthesized by dissolving 1M of $CuCl_2$ and 0.68mM of $Na_3C_6H_5O_7$ in water, according to the protocol suggested by *Zhou et al.* [8]. Our approach considered the systematic investigation of the effect of the sulfide precursor (Na_2S or TGA), temperature and reaction time. The Cu solution in citrate media was contacted with a 1M solution of Na_2S or TGA and heated in a microwave furnace Mars Xtraction, CEM at 400W at 60 °C. The heated suspension was quenched in ice to interrupt the progress of the reaction. Solids were recovered by coagulated with 2-propanol, centrifuged, dried at room temperature and characterized.

D.   Materials Characterization

The crystalline structures of the QDs were determined using a Siemens D500 x-ray diffractometer (XRD) using the Cu-K$_α$ radiation. The average crystallite size was estimated by using the Scherrer's equation. Samples were examined using a JEOL JEM-200FS high resolution transmission electron micrscope (HRTEM). UV–VIS absorption spectra were recorded using a UV-Vis DU 800 spectrophotometer. The emission spectra were obtained using a RF 5301 PC Spectrofluorometer Shimadzu at an excitation wavelength of 410 nm. All the absorption and emission spectra were recorded at room temperature within wavelength ranging from 200 to 800nm.

III. RESULTS AND DISCUSSION
A. CdSe QDs
Physical Characterization

XRD analyses were conducted on samples synthesized with a citrate/Cd mole ratio of 1.0 and a constant mole ratio of Se/Cd. The difractogram is shown in Fig. 1(a) and evidenced the formation of hexagonal nanocrystalline CdSe (Powder diffraction card 99-101-0836). The broadening of the peaks reflects the small size of the CdSe crystals; the crystallite size was roughly estimated to 8nm by using the Scherrer's equation. The lattice parameters were calculated to a=4.653Å and c=7.357Å. Fig. 1(b) shows a TEM image of isolated CdSe QDs with observed sizes in the 4 to 6 nm range. X-ray energy dispersive spectrometry confirmed the crystals to be composed exclusively of Cd and Se.

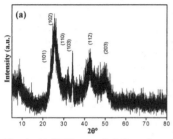 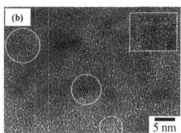

Fig. 1: (a) XRD recorded from the solid obtained using a citrate/Cd mole ratio of 1 at room temperature. All main peaks correspond to CdSe. (b) TEM image showing isolated CdSe QDs.

UV-VIS Analyses

Fig. 2(a) shows the UV-VIS spectra recorded from the CdSe QDs synthesized at different citrate/Cd mole ratios. The position of the exciton peak was slightly red-shifted when the citrate/Cd mole ratio was increased. This shift would suggest some growth of the CdSe nanocrystals promoted by the larger availability of citrate species in solution. The band gaps were estimated using Tauc's equation [10]. The estimated values varied between 2.10eV and 1.99eV. These values can be considered an evidence of the quantum confinement effect in CdSe QDs (band gap for bulk CdSe is 1.74eV). Based on these preliminary results, a citrate/Cd ratio of 1.0 was selected for the subsequent experimental work.

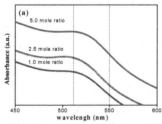 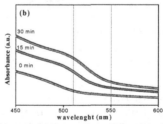

Fig. 2: UV-VIS spectra recorded from CdSe samples: (a) Different citrate/Cd mole ratios with a fixed reaction time of 30 min. (b) Different reaction times with a fixed citrate/Cd mole ratio of 1.

Fig. 2(b) shows the UV-VIS spectra of the selected citrate/CdSe ratio of 1.0 synthesized at 60°C but with different reaction times. The UV-VIS data evidenced the red-shift in the exciton peak at larger citrate/Cd mole ratios. The band gap energy values were estimated to 2.26, 2.20 and 2.05 eV, for 0, 15 and 30 min. of reaction, respectively. The large band gap energy value of 2.26eV suggests a smaller crystal size.

Photoluminescence Analyses
Photoluminescence spectra were recorded with an excitation wavelength of 410nm for samples synthesized at reaction temperatures of 25°C and 60°C. The synthesis at 60°C was carried out in a microwave furnace as described in the experimental section. Fig. 3(a) shows the PL spectra of those samples synthesized for 15 min. at 25°C and different citrate/Cd mole ratios. On a general basis, the emission peak was red-shifted at citrate/Cd mole ratios of 2.5 and 5.0 with respect to the spectrum obtained at a mole ratio of 1. This red-shift of the luminescence peak also suggests a promotion of the crystal growth, in good agreement with UV-VIS observations. In turn, the systematic drop in the luminescence intensity by increasing the citrate amounts can be attributed to the quenching effect by the promoted formation of a Cd-citrate layer on the surface of the QDs. The red-shift in the emission peak was also observed by prolonging the reaction times in CdSe nanocrystals (Fig. 3(b)). In this case, the observed shift is attributed to the promotion of crystal growth by prolonging the reaction time. A very sharp and intense luminescence signal was observed in the samples generated right after the addition of the reacting solutions (i.e. a reaction time of 0 min), which evidenced the very fast formation of CdSe nanocrystals. However, the luminescence intensity started to quench at longer reaction periods. The high intensity in luminescence can be related to the favored formation of electron-holes pairs at the beginning of the reaction; as the reaction progresses, more stable species would form making the availability of the electron holes pair to decrease [2,11]. The probable formation of $Cd(OH)_2$ during the reaction stage, stable at alkaline conditions, could also explain the observed quenching in the luminescence.

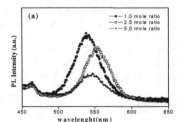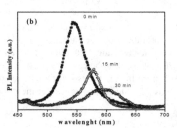

Fig 3: PL spectra from CdSe QDs synthesized with: (a) Different citrate/Cd mole ratios at room temperature, and 15 min. of reaction time; (b) A citrate/Cd mole ratio of 1, different reaction times and at 60° C.

B. CuS nanoparticles
Physical Characterization
XRD diffractograms for the CuS powders synthesized in the presence of Na_2S and using a hot plate or a microwave furnace as heating systems are shown in Fig. 4(a) and (b), respectively. The noisy pattern and the broad peaks in (a), suggests the formation of poorly crystalline solids. The detectable peaks could be assigned to the (110) and (108) planes of the hexagonal covellite type

CuS structure [12,13]. Although still broad, the diffraction peaks of Fig. 4(b) suggest the improved crystallinity in the CuS NPs synthesized under microwave-heating conditions. The expected high heating efficiency that can be attained in a microwave furnace can explain the enhancement in the crystallinity of the corresponding sample.

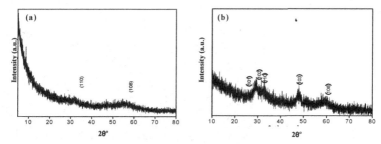

Fig. 4: XRD diffractograms recorded from CuS NPs synthesized using Na_2S as a source of sulfide ions heated up to 90°C with a hot plate (a), and with a microwave oven (b).

UV-VIS Analyses

Fig. 5(a) shows the absorption spectra of the CuS sample synthesized at 90°C using a hot plate or a microwave oven. Na_2S was used as the sulfide source in both tests. The strong and well defined absorption peaks, centered on 970-980nm evidenced the formation of CuS [8,14]. Fig. 5(b) shows the PL spectra for the samples microwave-heated up to 90°C but using Na_2S or TGA as the source of sulfide species. The comparison of the peaks intensity evidenced the more favorable formation of CuS when Na_2S was used; apparently, 90°C was not high enough to promote the thiol thermal decomposition and subsequent generation of free sulfide species in aqueous phase.

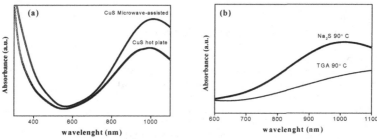

Fig. 5: UV-VIS spectra recorded from: (a) CuS synthesized in presence of Na_2S and heated using a hot plate or a microwave oven; (b) CuS synthesized in the presence of TGA using microwave oven. The reaction temperature was 90°C in all cases.

As an attempt to confirm our previous hypothesis regarding the need of higher temperatures to promote the sulfide formation, a different set of experiments were carried out in the temperature range between 90-120°C under microwave-heating conditions. TGA was used as the source of sulfide ions and the reaction time was fixed at 20 min. The PL spectra in Fig. 6 confirmed our previous interpretation. The increase in the reaction temperature favored the development of the CuS phase as suggested by the remarkable and systematic increase in the absorption peak centered on 1000nm. As suggested, the rise in temperature should have promoted the thermal decomposition of the thiol structure and consequent generation of free sulfide ions, as required by the formation of CuS [13,14]. A TEM image showing isolated CuS NPs is displayed in (b).

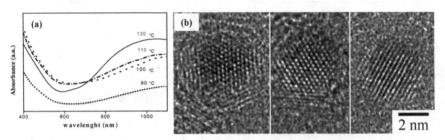

Fig. 6: (a) UV-VIS spectra of CuS synthesized in the presence of TGA at different reaction temperatures under microwave-heating conditions. b) TEM image of isolated CuS NPs. The crystals sizes were in the 3-5nm range.

IV. CONCLUSIONS

The systematic investigation on the formation of CdSe QDs in presence of citrate species revealed that a citrate/Cd mole ratio 1.0 and microwave-heating conditions (60°C) were conducive to suitable structural and optical properties in the as-synthesized QDs. In turn, the structural and optical characterization of the powders synthesized under microwave heating conditions confirmed the formation of CuS NPs by using Na_2S or thioglicolic acid as the source of sulfide ions. The thermal decomposition of the thiol species, and consequently the formation of the CuS phase, was promoted at higher temperatures.

ACKNOLEDGMENTS
This work is supported by NSF-CREST grant N°. HRD0833112.

REFERENCES
[1] P.P. Ingole et. al, *Materials Science and Engineering B*, **168**, 60-65, (2010).
[2] M.P. Melacon et. al, *J. Am. Chem. Soc.*, vol. **44**, 10, 947, (2011).
[3] W. Dong et.al, *Spectromic Acta Part A*, **78**, 537, (2011).
[4] Dethlefsen et. al. *Nano letters*, **11**, 1964, (2011).
[5] F. Masashi et. al. *J. Phys. Chem. C*, **113**, 38, (2009).
[6] Y.J. Bao et al. *Chinese Chemical Letters*, **22**, 843–846, (2011)
[7] Kloepfer et al. *Journal of Physical Chemistry B*, **109**, 9996, (2005).
[8] Zhou et. al, *J. Am. Chem Soc.*, **132**, 15351, (2010)
[9] C.M. Hessel, *Nano Lett.*, **11**, 947, (2011).
[10] A. Slav. Digest J. *Nanomaterials and Biostructures*, **6**, No 3, 915–920, (2011).
[11] Q. Tian et.al, *J. Am. Chem. Soc*, **5**, 9761, (2011).
[12] W.W. Yu et. al, *Biochemical and Biophysical Research*, **348**, 781, (2006).
[13] Vider et. al, Nano Lett., **9**, 442-448, (2009)
[14] T. Thanh, Green, L.A. *Nano Today*, **5**, 213-230, (2010)

AUTHOR INDEX

SUBJECT INDEX

Printed in the United States
by Baker & Taylor Publisher Services